Radionics Book 5: Delawarr Conversion Rates

Compiled by Peter J. Kelly

Edited by Robert Witney, Cathie Jordan, and Ed Kelly

Version: 2024

DISCLAIMER: Publication and distribution of the ideas, theories, concepts, descriptions, processes or instructions published in this book in no way represents or constitutes a claim that any product, system, or device manufactured or offered by Kelly Research Technologies, Inc., nor any other scalar, radionic, psychotronic, or free energy device, can have a positive, negative or any other impact on human health, or is in any way a replacement for or supplement to the advice or treatment provided by state and federally licensed medical and other commonly recognized allied health care professionals.

Kelly Research Technologies' radionic instruments and accessories are intended for agricultural research only and can in no way treat any medical condition or prevent any human medical condition from occurring. All radionic rates, frequencies, and other patterns of information are compiled for historical, energetic, and/or experimental agricultural research only and are not intended for use with humans at any times.

More specifically, Federal law states that only a medical doctor and/or prescription drug is/are able to diagnose, treat, prevent, or cure disease. All information and all correspondence verbally, via email, brochure, letter or otherwise does in no way make any claims to diagnose, treat, prevent or cure any ailment, affliction, suffering, or disease that any person(s) have had, currently have, or will have in the future. It is always recommended to discuss with your medical doctor prior to making any changes in diet, nutrition, supplementation and exercise.

If you are in need of medical or mental health care, please contact a licensed professional!

ISBN 978-1-7348707-5-6

Table of Contents

KRT
KELLY RESEARCH TECHNOLOGIES
www.kellyresearchtech.com

Statement of Mission

To distribute top quality technology, resources, and information to the global community of radionic researchers.

About Us

Located in Lakemont, Georgia, Kelly Research Technologies was incorporated in 2006 in order to sustain and advance the radionic technology developed by Peter J. Kelly from the early 1970's until his untimely passage in 2004. Core among his contributions was the design of a two-bank radionic instrument that integrated the Hieronymus design for variable plate capacitance signal tuning with the knowledge of solid state amplification he'd acquired while in service to the nation in the U.S. Navy. As a founding member of the U.S. Pyschotronics Association in 1974, there were many opportunities for friends and colleagues to try the new instrument. The "Kelly Instrument" would be built for friends, then friends of friends, then in 1978 the first one was sold to a complete stranger – the first of thousands that would choose this sensitive, reliable, and easy-to-operate radionic device.

Today, Kelly Research Technologies manufactures and distributes not only the famous Kelly Personal Instrument, but also a complete line of radionic instruments designed to meet the needs of both the advanced researcher and the busy farmer. In all Kelly instruments, the combination of variable capacitance tuning and solid state amplification allows determination of scalar frequencies and assessment of their intensities through the focused intent of the trained operator, as well as transmission of in-phase and reverse phase scalar frequencies. With thousands of two-dial radionic rates available, Kelly radionic instruments are ideal for all aspects of radionic agriculture, as well as many types of amplified dowsing.

Our 30-day, no-hassle, money-back guarantee ensures that every customer has a pleasant, professional experience. Whatever your reasons, simply return the item to us in like-new condition and receive a full refund of the purchase price. Your only risk is the price of postage. Find out why we've had a great reputation in the radionics community since 1978!

www.kellyresearchtech.com

Foreword to the 2024 Edition

Radionic rates are one of the foundational pillars of any system of live-signal radionic instrument. These frequencies are the strongest resonance points that have been identified for the many samples, reagents, and intentions that have been studied by the hardware inventors, the teachers, and the countless researchers who use radionics daily around the home and farm.

Kelly Research Technologies has been developing and compiling radionic rates since *before* the launch of the company in 1978. A crowning achievement was translation of the rate book for the multi-dial De La Warr instrument to the two-dial system popularized by T. Galen Hieronymus, which required setting, measuring, and recording nearly 1,500 rates on both types of instruments by KRT founder Peter J. Kelly. Another milestone was the 1983 publishing of *Psychotronics Book 2* by Peter Kelly, a collection of historical and new rates. This book was only available for a few years before it was banned by U. S. Food and Drug Administration because of my dad's naïve use of forbidden terms like "diagnose", "treat", and "cure" throughout. In 1998 the information was re-released as the *Electronic Rate Book*: a computer database with the offending terminology removed and the ability to search and sort the rates as never before. In 2011 the E-Rate Book was totally overhauled, the ever-growing database totaling more than 13,800 rates, the ability to create worksheets and reports on the fly, as well as access and print from a library of 28 pre-formatted worksheets organized on a variety of themes and processes. One unexpected benefit that came from our overhaul of the E-Rate Book software was that the process of organizing the information into book format was suddenly a realistic goal for our tiny staff of three people. In 2021, KRT launched a web-based version of the Electronic Rate Book at www.kellyratebook.com. Completely free for all users, this resource opened the door to everyone that had access to a web browser.

Despite the continuous enthusiasm for electronic searching and sorting of radionic rates, there has always remained a demand for old fashioned printed rate books. Since 2012, we've published hundreds of copies of *KRT Radionics Book 3: Radionic Rates - Alpha*, an alphabetical sort of the entire KRT database. For this reason, we were proud to present the 2022 edition of *Book 3*, which now includes more than 14,450 radionic rates. Likewise, we at last published the long-awaited sister volume, *KRT Radionics Book 4: Radionic Rates - Numerical*, which presents the same database of information sorted numerically by left-hand dial and then right-hand dial. And finally in 2024, a new version of Book 5: Delawarr Conversion Rates. The 2024 edition has been reorganized to improve clarity and alphabetical and numerical sorts of the database have been added. We hope the community will benefit from the work represented and shared in this volume.

Ed Kelly, President,
Kelly Research Technologies, Inc.
Lakemont, Georgia

Introduction to the 1983 Edition

Ahhh yes, those wonderful Delawarr Rates.

George & Marjorie de la Warr

The Delawarr Rates were derived from the rates Ruth Drown develeoped for specific information used in her research. George de la Warr, one of the pioneers in creating modern radionics, needed to get down to a very minute level because of the relative ineffectiveness of the earlier instrumentation being used for affecting the energetic levels of conditions in the various locations. A typical Delawarr Instrument rate might have as many as 10 digits for each aspect whether it indicated a Condition or a Location. There were as many as 20 dials that needed to be turned and tuned to get the same accuracy as we seem to be able to get with two dials on Variable Capacitor type experimental instrumentation. The procedure for using up to 20 dials was tedious at best. Yet the information they achieved is still invaluable to researchers.

A Delawarr Instrument

In order to make available their research information for those doing research with the Variable Capacitor type instrumentation, we set up a very specific procedure to convert the information. To do this, we needed a Delawarr Instrument and connected an output to it, which we fed to the the input of the Experimental Agricultural Analyzer with variable capacitor tuning. Then the Delawarr Information was set on the Delawarr machine while a scan was done on the Variable Capacitor tuner for the equivalent "two-dial" information.

This little volume represents several months of work scanning, checking, and cross-checking for relative accuracy and as such became an important early part of our understanding of how this "energy-as-information" works. This information has become a standard in Variable Capacitor Tuning Research over the last 20 years. Enjoy the fruits of our labor in your own personal research.

Peter J. Kelly, President
Interdimensional Sciences, Inc.
Lakemont, Georgia

How to Use this Book

After the introductory sections, this book is divided into three parts: Energetic Anatomy, an alphabetical sort of the database, and a numerical sort of the database. Suggestions and tips for their use are as follows:

How to Use the Energetic Anatomy Database

The Energetic Anatomy Database is organized by major anatomical systems of the human and animal body, though its pivotal to understand that the radionic rates and other information provided refer only to their energetic counterparts, with no medical bearing on physical anatomy, health, or wellness. Any decisions of a medical or mental health nature should be made under the guidance and advice of a licensed professional.

Energetic Analysis and Balancing: The rates provided allow in-depth analysis into the relative balance of energetic anatomy of the subject.

- With the exception of the section "Energetic Diseases and Disorders", all rates tested should ideally present an Intensity reading equal to the Intensity measured for General Vitality [9.00-49.00]. For any rates presenting an Intensity reading below General Vitality, the user may make a balancing broadcast back to the subject using the instrument-specific techniques described in *KRT Radionics Book 2: Applied Radionics*, or the manuals of comparable two-dial radionic instruments.

  Note: We do not recommend trying to lower the energetic intensity of any Patterns of Information found to exceed General Vitality. Rather, we prefer to raise the entire energy system upward until all elements are in balance.

- For those rates and patterns described in the section "Energetic Diseases and Disorders", all rates tested should ideally present an Intensity reading equal to zero. For any rates presenting an Intensity reading above zero, the user may make a balancing broadcast back to the subject. Because the patterns of information for diseases and disorders are not hard coded into the DNA of the subject, the balancing broadcast should lower the measured intensity to or toward zero.

- Some Patterns of Information will also include additional rates or suggestions for things to check or balancing rates to try. Users may also use supporting rates, reagents, affirmations, and hertzian frequencies to help raise or lower intensities, as with any other radionic broadcast.

Multiple Rates: Some Patterns of Information in the Energetic Anatomy Database will include multiple rates. These are *all* single-bank rates, meaning that each rate may be used individually for both analysis and balancing, though left- and right-hand dial settings must always be used together as printed. You can no more separate a left dial setting from its right dial counterpart than you could split a telephone number in half and still expect to reach the person you are trying to call! That said, users may also simultaneously analyze and balance using two or more of the rates that are listed with a particular pattern of information. Given that every radionic rate is a frequency or tone that is then converted into a scalar field, using multiple rates simultaneously can create a more complex chord that can bring forth a more complete picture of the situation at hand and/or yield a bigger movement of energy.

Dig Deeper to Find and Resolve Energy Problems: The in-depth anatomical descriptions found here can allow the user to dig much more deeply than "top level" rates that cover the main organ or system. For example, an energetic stomach problem can be checked using the rate [0.00-32.00], but if the real issue is an energy blockage of the stomach's gastric lymphatic nodules, a much quicker and more lasting resolution can be had by balancing that specific rate. When in doubt, check for intensities balanced with General Vitality for *all* the subsystems connected to an ailing energetic organ or system.

Repeating Descriptors are Specific to that Section: Careful readers will notice that many energetic organs and systems will have terms that repeat from section to section, such as arteries, lymphatics, nerves, and so on. In each case, the rates provided refer to the arteries, lymphatics, and nerves specific to that energetic organ or system, which have been found to resonate at their own rates.

How to Use the Alphabetically-Sorted Rate Database

Header: The top right corner of each page features the first three letters of the first Pattern of Information listed on that page, allowing the user to quickly flip to an entry. After the introductory pages, the page number is found centered at the top of each page.

Patterns of Information: Every radionic rate in this book was originally identified when a Peter Kelly used a two-dial radionics tuner to locate a pattern of energetic resonance when an attached Delawarr instrument was tuned to a specific setting. This unique resonant or dissonant relationship with the original sample provides the basis for the researcher's efforts to alter the energetic balance in the crop, soil, or barn. An alphabetic listing of this information allows the users to quickly locate rates that other users have identified as having resonance in similar situations and may provide clues in the search for the cause behind the cause in a given situation.

However, it is also important to remember that the names given to these rates by their discoverers may or may not apply to the specific situation being considered by the current researcher. Whether or not a particular pattern of information is actually applicable to that situation must be tested by that operator at that time. Always test for desirability, suitability, and permission before broadcasting using the Intensity Dial on your instrument.

Immune System, Molecule, Cytokines, May be causative factor in Wasting Types Diseases	34.00	54.25
Implant Mechanism, (1) Negative Entity	36.50 21.00	36.50 05.50
Mold, Rinkel Mold Mix A	-97.50	23.25
Moles & Warts	26.00	32.50

Figure 3: Examples of rates from this book.

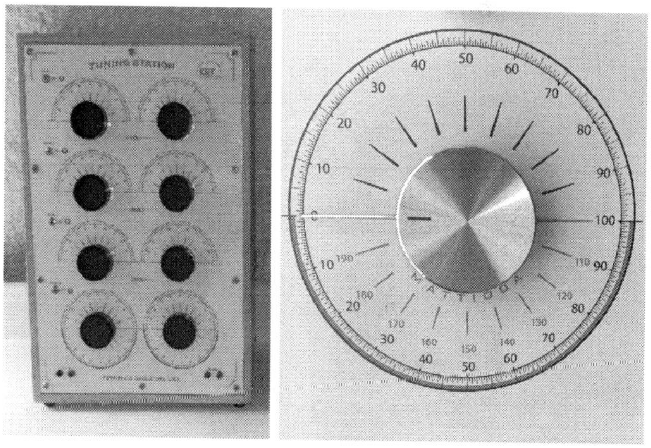

Figure 4: A KRT Tuning Station and 360° "round dial" rates.

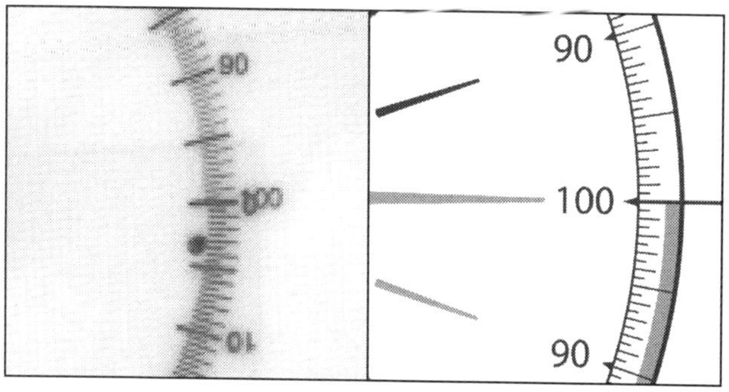

Non-KRT Instrument:
100.00 = 0.00

KRT Instrument:
100.00 = 100.00

Figure 5: Comparing 360° rate dials.

Rates: All rates are presented in pairs with two digits of decimal accuracy for each rate. Remember that fractional amounts to the right of the decimal point are the fractions of the space between tick marks on the rate dial. The shortest tick mark shown on the KRT rate dial is 0.50 (see Figure 5). The examples in Figure 3 will be used to illustrate how radionics rates are presented in this book:

Single Bank Rates – All of the Delawarr Conversion Rates require only a single rate bank to use. *Moles & Warts* is a typical single-bank rate and would be set on the instrument by turning the left dial until the needle was set to 26.00, setting the right-hand dial to 32.50, then activate that bank by turning on the switch. Note that both the left and right hand dials must be set in order to achieve the desired information or impact. The rate for *Immune System, Molecule, Cytokines* would be utilized in the same manner.

Multiple Bank Rates – None of the Delawarr Conversion Rates require more than one bank, however, the operator can use multiple banks to analyze and/or balance combinations of rates. Likewise new rates may be cold scanned that personalize the Delawarr rates to the subject, or otherwise improve their effectiveness. In this case, set the Delawarr rate of interest on one bank, then cold scan on the additional bank(s). Use the Intensity dial to verify that cold scanned rates are increasing the strength of the patterns of energy you wish to make stronger, or decreasing the strength of the patterns of energy you wish to diminish.

Round Dial Rates – Some radionic instruments are equipped with 360° rate dials that allow the researcher to explore the other half of the waveform when identifying a pattern of energetic resonance in a sample (see Figure 4). With *Mold, Rinkel Mold* the operator will set -97.50 in the "green field" area of the dial on a Kelly instrument, such as a Seeker, Workstation, or Tuning Station. Note that the minus sign only applies to the one rate. The right-hand dial setting of 23.25 will be set in the black-only upper part of the dial. While no Delawarr rates were found in the "green field" area of the dial, complementary green field rates may by scanned by the user.

Note to users of non-Kelly instruments equipped with 360° rate dials. Be sure to check the numerical arrangement on the lower part of your 360-degree rate dials before using a rate from this book. On some instruments the number "100" is followed immediately by the number "-1" rather than "-99" as shown on Kelly instruments. Because all round dial rates in the Kelly database are configured for our own instruments, users of instruments configured the other way must be sure to convert the rate for use with their instrument by subtracting the number from -100.00. For example, for *Mold, Rinkel Mold:* -100.00 – (-97.50) = -2.50, the rate set on the "100=0" style instruments made by other vendors.

Energetic Processes

This field of the database often provides supplementary information intended to help the radionic researcher use the listed Pattern of Information successfully. More specific information may include:

- **Analysis-only Rates:** Some rates are intended to be used in analysis mode only. If so, this will be noted and, usually, the correct rate for balancing will be provided.
- **B/R:** Simply means "balance on rate" - use the listed rate for the broadcast.
- **Complementary Rates:** This field will often include rates that can be tested at the same time or in series with the primary rate listed. Always test for applicability.
- **Frequencies:** Abbreviated as "Freq", the frequencies listed in the Energetic Processes field are non-radionic waveforms listed in cycles per second, or Hertz. Some of these would be audible tones if played from a signal generator equipped with a speaker, while others would be above or below the range of typical human hearing. Frequencies listed should be treated as electronic reagents, which is to say that they can be used concurrently during balancing broadcasts to increase impact or effectiveness.

Remember that all information captured in the "Energetic Processes" field was developed/discovered by an individual working on a specific set of challenges at a particular time. *Always test to verify* that the information presented is relevant and appropriate for the research you are doing at this time.

Note on using reagents: Whether the reagent is an electronic frequency (added through the "SIGNAL IN" port) or a sample of a color, essential oil, or supplement (placed in the input well), every reagent must be tested to ensure that the desired impact is being achieved. The primary question for the researcher should be,

Does this reagent increase or decrease the vitality of my research subject?"

Does adding the reagent or the frequency make the General Vitality or part of the organism go up or down? If it is a pathological pattern of information, does the reagent/frequency make it weaker or stronger? Testing the applicability of a reagent only requires taking a new reading on the Intensity dial to verify that the desired outcome will be achieved.

How to Use the Numerically-Sorted Rate Database

Background: Cold scanning of radionic rates is a valuable tool for any radionic researcher. While novice operators tend to focus on lists and worksheets of known rates to address energetic problems that they can identify, more experienced operators will also utilize cold scanning to find those rates that can most effectively add energy and/or clear blockages. They can simply cold scan for the best radionic rate for the situation at hand, then test to ensure the scanned rate will raise the vitalities of those energies they want to increase or lower those negative patterns they wish to decrease. By trusting their own skills, they are set free from the need to know the labels for the issues that are preventing their projects from reaching their highest potentials.

Typically, the operator places a sample or a witness in the well and searches the tuning dials for the strongest resonance points associated with that sample one or more banks on the instrument. The process may also be driven by focused intent, with the operator focused on questions or statements that relate to the sample or witness, such as "What does this crop need most at this time?" or "Remove all obstacles associated with the success of my project." Likewise, the cold scanned rate may simply be a "personalizing" or "supporting" rate for harmonizing a broadcast of known rates with the sample or witness on hand.

In many cases, the rate that is cold scanned is treated as a "single serving" rate that is used once and forgotten. After all, "what we need most" will change from day to day and can always be scanned again from scratch as needed. However, with important projects and persistent issues, the cold scanned rates may be explored more deeply for clues regarding the "cause behind the cause" that is preventing a desired manifestation from occurring or helping a negative pattern of energy to recur. This is where a numerical sort of the radionic rate database can be invaluable.

Application: Having cold scanned for a radionic rate as described in the preceding chapter, the operator can use the numerical section of *KRT Radionics Book 5* as a cross reference to see if:

1. The cold scanned rate can be exactly matched to one within the database:

   If the rate scanned by the operator is a perfect match for one in the database, its very possible that there is additional information that may be inferred from either the label/name ("Patterns of Information") or even directly applied from the supporting notes ("Energetic Processes").

   For example, if the scanned rate turns out to be a vitamin, then that vitamin will probably be an excellent reagent to add to the sample well. Or the need for the energy of that vitamin might speak to a blockage of energy to the organ that would most directly benefit from a greater availability of that substance.

2. The cold scanned rate can be partially matched to one within the database:

   Much more common is that one or other of the scanned dials is a match for something in the database, while the other number is not. In some cases, this information may still be useful. In the classic single bank radionic instrument system devised by T. Galen Hieronymus, the left dial was assigned to represent energetic conditions, while the right dial was assigned to parts of the anatomy or the energy system.

   For example, the condition "Viral Diseases" has the rate "2.00-0.00" and "Colds, Flu" has the listing of "3.00-0.00". According to Hieronymus, these rates would represent "Viruses in an unspecified location", or "Flu in an unspecified location", respectively. Likewise, the rate "0.00-25.00" is associated with the male prostate gland, while "0.00-44.00" is associated the energy of the lymph nodes. In fact, one might notice that most of the rates found on the early pages of the numerical database have a zero in the left-hand position and a number in the right-hand position and are associated with parts of the energetic body. Using this logic, it is sometimes useful to check the left-hand dial and right-hand dial independently of one another to see if there is a useful clue as to the root source of the problem the operator is trying to resolve.

Example: If the cold scanned rate was "22.00-73.00", we will not find an exact match for that rate in the database. But we can check "22.00-0.00" and we find six partial matches that follow the form "22.00-0.00". (Note that the first two entries are four-dial rates that would be set on two banks at the same time.) Do any of these make any sense or apply to the problem at hand? This is a question that the operator must answer based on their knowledge of the situation! If the project is related to the energetic health of livestock, then its likely they have found a relevant clue.

Chemical, A T Pase Needed For Chemistry of Energy	22.00 06.00	00.00 02.00
Amino Acid, Serotonin; Brain Neurotransmitter	22.00 00.00	00.00 00.25
Animal, Diseases, Swine, Stress	22.00	00.00
Gastrointestinal Disorders, Colon, Hypertonicity	22.00	00.00
Gynecologic, Obstetric, Pediatric, Genetic Disorders, Child Disorders, Hyperactivity	22.00	00.00
Brain, Fore, Hypothalamus, Use For	22.00	00.00
Fungus, Schimmelpilz II	22.00	00.25

Similarly, we can look at "0.00-73.00" and we only see one match: "Respiratory, Lungs, Upper". Again, whether this is useful information or makes any sense in the project at hand must be decided by the operator. If the issue is a common cold, then its possible that the cold scan is indicating that energy tune up for the respiratory system would be in order.

Throat, Larynx, Glottis In	00.00	71.50
Urinary, Kidneys, Tubules, Right	00.00	72.00
Respiratory, Lungs, Upper	00.00	73.00
Respiratory, Lungs, Lower	00.00	74.00
Colon, Diverticulum	00.00	75.00

In this way, the numerical sort of the radionic rate database can provide clues to pursue when searching for the root causes that lie behind the surface problem the operator is trying to resolve.

Note: *If a cold scanned rate does not make sense when broken down in this way, it does not mean that the rate is invalid or unusable.* The bottom line is always to check to see if the new rate raises the vitalities of those energies they wish to increase or lowers those negative patterns they wish to decrease.

Does this cold scanned rate increase or decrease the vitality of my research subject?"

Specifically, the most basic test is:

1. Place the witness or sample for the subject in the input well.
2. Set one bank on the instrument to General Vitality (9.00-49.00).
3. Take a reading with the Intensity Dial
4. Set the new, cold scanned rate on a different bank on the instrument.
5. With both banks turned on, take a new reading with the Intensity Dial.

If the combination of the GV + the cold scanned rate is higher than GV by itself, then the cold scanned rate is raising General Vitality of the plant or animal. Thanks to the ease and simplicity of this test, even the newest beginner may test their work and feel confident in their new cold scanning abilities.

Cell Structure

Cells and the products of cells comprise all the tissues of the body. All functional activities of the body are carried on by the cells, and the structure and form of a cell is closely correlated with its functions. New cells come about by the division of existing cells. Growth and development of results from the increased number of cells and the differentiation of cells into different type of tissues.

Protoplasm: a thick, viscous colloidal substance that constitutes the physical basis of all living activities	50.00-46.60
Cyocentrum Attraction Spheres: minute bodies in cytoplasm of cell close to the nucleus	64.25-73.00
Centrioles: minute organelles consisting of hollow cylinders closed at one at one end and open at the other. Found in the cell center or attraction sphere of a cell	59.00-57.10
Cytoplasm: the protoplasm of a cell outside the nucleus	53.20-26.00
Chromophil Granules: one of two types of cells present in pars distalis of the pituitary gland. It is considered a secretory cell	45.00-60.50
Golgi's Apparatus or Network: a lamellar membranous structure near the nucleus of almost all cells. It contains curved, parallel series of flattened saccules that are often expanded at their ends	75.50-75.50
Hyaloplasm: the fluid portion of the protoplasm	55.75-64.50
Mitochondria: in brown fat cells, slender microscopic filaments or rods and are the source of the energy in the cell	54.50-39.00
Ton Fibrils:	69.75-55.50
Nucleus: the membrane-enclosed organelle within a cell that contains the chromosomes	47.60-66.00
Chromatin Granules: a deeply staining substance in the nucleus that that contains the genetic material	64.10-65.30
Chromosomes: a structure in the nucleus of the cell containing the linear threads of DNA – the blueprints of the cells	55.50-52.00
X-Chromosomes: one of a pair of female-determining chromosomes present in the somatic cells of all human females	60.75-56.00
Y-Chromosomes: one of a pair of male-determining chromosomes present in the somatic cells of all human males	50.00-44.50
Genes: the basic unit of heredity	54.50-56.20
Nuclear Membrane: membrane enclosing the nucleus	47.50-49.00
Nucleolus: a spherical body within the cell nucleus	50.50-31.50

Central Nervous System

The central nervous system [5.00-62.00] includes the brain, spinal cord, the nerves, and end-organs that control voluntary acts. It includes the part of the brain governing consciousness and mental activities; parts of brain, spinal cord, and their sensory and motor nerve fibers controlling skeletal muscles, plus end organs of the body wall.

Autonomic Nervous System: 40.40-42.50
the part of the nervous system that controls involuntary bodily functions

- Parasympathetic System: the craniosacral division of the autonomic system 50.00-42.00
  - Nerve Fiber: form the major portion of the white matter of the brain, spinal cord, and all nerves 57.00-62.50
- Sympathetic System: 5.00-60.00, 5.00-61.50
  the thoracolumbar division of the autonomic system
  - Ganglia, Ganglion: masses of nervous tissue composed principally of nerve-cell bodies found outside the brain or spinal cord 35.00-37.50
  - Plexuses: networks of nerves 45.25-60.10

Brain Areas:

- Association Area: 53.50-80.75
  - Frontal: 0.00-5.00, 0.00-82.75, 37.00-26.00
    With depression, check need to increase melatonin, noradrenaline, or serotonin, or decrease dopamine
  - Post Antral: area of memory 41.60-29.60
    With depression, check need to increase noradrenalin or serotonin; or decrease dopamine
  - Auditory Area: temporal lobe 39.25-65.50
  - Balance & Equilibrium: cerebellar cortex 48.25-43.00
  - Coordination of Movement: cerebellar cortex 57.25-40.75
  - Post-Rolandic Area: posterior to the fissure of Rolando (sulcus centralis – the fissure dividing the frontal and parietal lobes of each hemisphere 23.50-25.00
    - Arms 40.40-35.40
    - Face 35.75-14.50
    - Legs 32.50-25.00
    - Trunk 80.50-63.75
  - Pre-Rolandic Area: 63.00-48.25
    - Arms 58.50-52.50
    - Eyes & Head 38.00-37.50
    - Face 52.50-28.50
    - Legs 40.00-53.50
    - Trunk 48.75-46.25

Sensory Area:

- Smell 82.20-60.50
- Taste 60.00-55.00
- Uncinate Gyrus 39.50-39.00

Sex Areas:

- Desire 96.25-15.00
- Endocrine Center 34.75-52.00
- Frustration: base, anterior auditory artery 69.50-71.00
- Function: 47.50-51.00
- Loss of Libido: 43.00-54.00

Sleep Center: 43.50-43.50

Speech, Broca's Area: area on the left side of the brain that contains the motor speech area and controls movements of the tongue, lips, and vocal cords. Loss of speech may follow hemorrhage in this area 42.75-47.25

Visual Area: Occipital 41.25-46.00

Visuo-Psychic Area: 45.50-45.00

Cranial Nerves: 33.50-35.75

Abducens: drawing away from the medial line of the body 37.50-38.00

Auditory: relating to the sense of hearing 41.75-29.50

Facial: relating to the face 49.50-33.75

Glossopharyngeal: relating to the tongue or pharynx 39.25-38.50

Hypoglossal: situated under the tongue 39.75-45.00

Oculomotor: relating to eye movements 43.60-43.50

Olfactory: relating to smell 28.40-28.50

Optic: relating to the eyes or eyesight 22.00-82.00, 44.50-18.50

Spinal Accessory: 27.00-33.40

Spinal or Peripheral Nerves: occurring away from the center 31.75-27.75

- Cervical: 8 pairs, region of the neck 1.00-21.00, 31.75-30.75
- Coccygeal: 1 pair, region of the coccyx 41.25-32.50
- Lumbar: 5 pair, region of the lower back or loins 35.50-18.50
- Sacral: 5 pairs, region of the sacrum 40.50-28.25
- Thoracic: 12 pairs, region of the chest or thorax 33.50-30.25

Trigeminal: 5th Cranial Nerve 43.00-42.00

Trochlear: relating to a trochlea (pulley) such as the certain joints 48.75-34.20

Vagus: 10th pair of cranial nerves arising in the medulla oblongata and providing parasympathetic innervation to the abdominal organs, esophagus, heart, larynx, and lungs. One of the most important nerves of the body. 39.50-27.50

Forebrain: 0.00-78.50, 57.75-87.50
With depression, check the need to increase enkephalins, norepinephrine, or serotonin; or decrease dopamine or noradrenaline.

Cerebral Cortex White Matter: 79.00-81.00
With depression, check the need to increase enkephalins, norepinephrine, or serotonin; or decrease dopamine or noradrenaline.

Cerebral Cortex: the connection between the hemispheres of the brain 50.00-38.00
With depression, check the need to increase enkephalins, norepinephrine, or serotonin; or decrease dopamine or melatonin.

Cowper's Glands: a pair of tubular glands beneath the bulb of the male urethra. 0.00- 2.00
Should equal General Vitality. Balance with [22.00-0.00] or [77.00-0.00].

Hypothalamus: lies beneath the thalamus and is laterally continuous with the subthalamic regions. Contains neurosecretions that are important to the control of metabolic functions such as water balance, sugar/fat metabolism regulation of body temperature, and releasing/inhibiting hormones 0.00-90.50

Nerves: 0.00-51.50, 53.00-50.50
Should equal General Vitality. Balance with [22.00-0.00] or [77.00-0.00].

Pineal, Pineal Body: 0.00-27.00, 16.00-28.00, 15.75-28.25
a gland-like structure in the brain located in the pocket near the splenium of the corpus collosum. *Often low with nervous conditions. With depression, check the need to increase melatonin, noradrenaline, or s erotonin; or decrease dopamine.*

Thalamus: 0.00-54.50, 52.50-40.00
All sensory stimuli except olfactory are received by the thalamus.
With depression, check the need to increase norepinephrine.

- Caudate Nucleus: 45.00-28.00
  With depression, check the need to increase melatonin, noradrenaline, or serotonin.
- Internal Capsule: 34.75-33.20
  With depression, check the need to increase norepinephrine.
- Lentiform Nucleus: 47.00-31.75
  With depression, check the need to decrease enkephalins or serotonin.
- Ventricle, Lateral: 61.50-62.00
- Ventricle, Third: 65.25-58.00

To Improve Memory, Nerves: 40.50-34.50
Balance with hippocampus [32.50-42.00] and acetylcholine [39.50-36.25]. With depression, check the need to increase enkephalins, norepinephrine, or serotonin; or decrease dopamine.

Hindbrain: main area of the brain 80.75-72.00

Cerebellum: area of emotions 0.00-43.25
With depression, check the need to decrease enkephalins.

- Dentate Nucleus: a cluster of neurons that has a serrated, tooth-like edge 57.25-39.75
- Fourth Ventricle: 0.00-4.50, 19.50-10.00
- Matter, Gray: grayish tissues with less myelinated nerve fibers 44.50-38.00
- Matter, White: any nervous structure composed of white medullated fibers 54.00-58.75
- Nerves, Left: 2.00-29.00
- Nerves, Right: 2.00-31.00
- Nerves, Epileptogenic Zone: 57.75-46.75
- Peduncle, Inferior: a band of fibers connecting spinal cord and medulla with the cerebellum 22.75-21.00
- Peduncle, Middle: a band of fibers connecting cerebellum with basilar portions of the pons 41.50-47.25
- Peduncle, Superior: a band of fibers connecting cerebellum with midbrain 21.50-21.25

Medulla Oblongata: enlarged portion of spinal cord in cranium. The lower portion of the brain stem. 28.00-13.00

- Olivary Nucleus: 57.00-83.50

Pons, Varolii, Fore, Medulla: a rounded eminence on the ventral surface of the brain stem. It lies between the medulla and cerebral peduncles. 0.00-15.00
With depression, check the need to decrease dopamine.

Midbrain: main area of the brain 30.25-30.50
With depression, check the need to decrease dopamine.

Cerebral Peduncles: a pair of white bundled from upper part of the pons to the cerebrum – constitutes the ventral portion of the midbrain 39.25-19.00

- Aqueduct of Midbrain: canal connecting third and fourth ventricles 41.75-16.50
- Basis Pedunculi: a band connecting parts of the brain 39.50-32.00
- Corpora Quadrigemina: the superior portion of the midbrain consisting of two pairs of rounded bodies, the superior and inferior colliculi 32.10-32.50
- Red Nucleus: large, oval, pigmented mass in midbrain that receives fibers from the cerebral cortex and cerebellum 43.25-38.25
- Substantia Nigra: black substance in a section of the crus cerebri 44.50-26.50
  With depression, check the need to increase melatonin or decrease serotonin.
- Tegmentum: dorsal portion of the cruri cerebri of the midbrain. It contains the red nucleus, nuclei, and roots of the ocular motor nerve. 37.75-35.00

Spinal Cord: the ovoid column of nervous tissue; all nerves to the trunk and limbs are issued from the spinal cord. It is the center of reflex action containing the conducting pathways to and from the brain.

Anterior Horns:	44.50-51.00
• Coccyx Balance:	40.50-22.00
• Extradural Space: unconnected with the dura mater	48.25-47.75
• Matter, Gray: grayish tissues with less myelinated nerve fibers	44.75-39.50
• Nerve Cells: a neuron, consisting of a cell body and its processes	73.20-46.20
• Nerve Fibers: elongated process of a neuron, usually the axon, concerned primarily with the condition of impulses. Nerve fibers make up most of the white matter of the brain, the spinal cord, and all nerves.	48.75-68.50
Meninges: the three membranes investing the spinal cord and brain, the dura mater (external), the arachnoid (middle), and pia mater (internal)	
• Arachnoid Villi:	3.25-46.25
• Cerebrospinal Fluid:	68.25-57.50
• Choroid Plexuses:	64.25-45.40
• Mater, Arachnoid (middle): thin, delicate, intermediate membrane that encloses the brain and spinal cord	48.75-40.40
• Mater, Dura (external): the outer membrane covering the spinal cord and brain	59.50-52.50
• Mater, Pia (internal): thin vascular membrane closely investing the brain and spinal cord, plus the proximal portions of the nerves	87.00-34.00
• Meninges of Brain and Spinal Cord: the three membranes Investing the spinal cord and brain	27.00-9.00, 8.00-92.00
• Nerves, Meninges:	0.00-50.50
Subarachnoid Space: space between the pia proper and arachnoid containing the cerebrospinal fluid	44.75-40.20
Subdural Space: space between the arachnoid and the dura mater	35.50-73.00
• Matter, White: any nervous structure composed of white nerve fibers	70.50-28.75
• Medullated Nerve Fibers: nerve fibers possessing a myelin or medullary sheath; a myelinated nerve fiber	59.00-39.00
- Anterior Column:	46.75-44.00
- Lateral Column:	48.50-38.00
- Posterior Column:	49.60-37.50

Circulatory System

The cardiovascular system, consisting of the heart, lymphatic system, and blood vessels, including the arteries, arterioles, capillaries, venules, veins, and sinuses.

Blood

Blood [0.00-19.00] is a highly specialized fluid called plasma, which consists of several types of cells that circulate through the heart [0.00-70.00], arteries [0.00-71.00], veins [0.00-69.00 or 0.00-97.00], and capillaries [66.75-44.25], carrying nourishment, electrolytes [86.50-26.00], hormones [53.50-58.00], vitamins, antibodies [6.00-72.00], heat, and oxygen [67.00-58.00] to the tissues, while taking away waste material and carbon dioxide [34.00-63.00]. The cellular constituents are: red blood cells [34.00-26.75], which carry respiratory gasses and give it its red color because they carry hemoglobin (an iron-containing protein that binds oxygen in the lungs and transports it to tissues in the body, white blood cells (leukocytes) that fight infection, and platelets (cell fragments that play an important role in the clotting of the blood). Anatomically, blood is considered a connective tissue from both its origin in the bones and its function.

Blood Corpuscles: the solid or cellular elements of the blood	14.00-36.20
Hemoglobin: the iron-containing pigment of the red blood cells whose Function is to carry oxygen from the lungs to the tissues	49.75-22.75
Lymph Corpuscles, Lymphocytes:	52.30-54.25
Plasma Cells: the liquid part of the lymph and the blood. *With depression check the need to decrease enkephalins, melatonin, serotonin*	55.60-61.75
Plasma Salts: *With depression check the need to increase norepinephrine or decrease enkephalins, melatonin, serotonin*	63.75-67.50
Platelets: small colorless bodies in circulating blood *With depression check the need to increase norepinephrine or decrease noradrenaline*	58.75-48.10
Red Corpuscles, Erythrocyte:	51.20-41.30
Reticulocyte: a red blood cell containing a network of granules or filaments representing an immature stage in development, normally comprising 1% of circulating red blood cells	35.00-29.75
White Corpuscles, Leukocytes:	
Basophil: cells or parts of cells that are readily stained with basic dyes	32.60-56.75
Eosinophil: A cell or cellular structure that stains readily with the acid stain Eosin. Specifically refers to a granular leukocyte. *If value higher than General Vitality, may indicate parasite(s) or allergies.*	70.00-75.00
Leukocytes: granulocytes process granules in their cytoplasm while Agranulocytes are those that lack granules	59.00-42.00 59.25-84.20
Lymphocytes: lymph cells or white blood corpuscles without cytoplasmic granules.	52.25-49.00

Macrophage: cells of the reticuloendothelial system having the ability to phagocytose particulate substances found in loose connective tissue and various organs of the body — 65.00-59.00

Monocyte: a large mononuclear leukocyte having more protoplasm than a lymphocyte — 23.75-25.50

Polymorphonuclear: possessing a nucleus with several lobes of fine strands — 37.25-56.75

Neutrophil: a cell that stains easily with neutral dyes — 9.50-69.00

Oxyphil: a cell that stains easily with acid dyes — 51.60-28.00

Hemostatic Agents: any drug, medicine, or blood component that serves to stop bleeding, such as Vitamin K, whole blood, or epinephrine applied locally.

Fibrin: a filamentous protein formed by the action of thrombin on fibrinogen — 40.00-32.40

Fibrinogen: a protein that helps form blood clots, high elevations

Indicate hidden risk of heart disease energy — 65.00-64.50

Indicate hidden risk of tissue fibrinogen energy — 61.50-69.00

Heparin: a polysaccharide isolated from lungs, liver, and other tissues — 65.75-59.00

Prothrombin: a chemical substance existing in circulating blood that interacts with calcium salts to produce Thrombin — 49.00-50.50

Thrombin: an enzyme formed in shed blood that reacts with soluble fibrinogen to convert it into fibrin, which forms the basis of a blood clot — 45.00-55.40

Thrombokinase: the tenth blood coagulation factor, factor x — 64.60-40.00

Thromboplastin: a substance found in both blood and tissues — 30.75-55.25

Plasma: liquid part of the lymph and of the blood — 31.20-19.00

Blood Vessels

Arteries: — 0.00-71.00, 68.00-71.00, 81.25-78.00

one of the vessels carrying oxygenated blood from the heart to the tissues

Tunica Adventitia - Outer Coat: layer in the outer wall of the tubular structure — 32.50-30.75

Tunica Intima - Internal Coat: layer in the inner wall of the tubular structure — 67.00-59.50

Tunica Media - Middle Coat: layer in the middle wall of the tubular structure — 28.75-29.50

Nerves: — 46.60-54.00

Peripheral Nerves: nerves that occur away from the center — 61.75-55.25

Perivascular Plexuses: the Splanchnic arteries — 70.25-55.25

Plexuses: a network of nerves — 58.40-76.20

Sympathetic Ganglia: deliver information to the body about stress and impending danger and are responsible for the "fight-or-flight" response. They contain 20,000 to 30,000 nerve cell bodies and are located close to and on either side of the spinal cord in long chains. — 39.50-84.00

Arterioles: a minute artery that leads into a capillary	73.25-40.75
Arteriole Capillaries: the terminal branches of the arterioles or metarterioles	32.75-41.60
Capillaries: blood vessels that connect the smallest arteries with smallest veins	66.75-44.25
Cells of Capillary Wall:	39.00-50.00
Cells of Envelope: the cell membrane and cell wall plus an outer membrane	59.25-52.50
Veins:	0.00-69.00, 0.00-97.00, 50.00-40.75
blood vessels that carry dark red unaerated blood back to the heart	
Tunica Adventitia - Outer Coat: layer in the outer wall of the tubular structure	61.00-73.10
Tunica Intima - Internal Coat: layer in the inner wall of the tubular structure	40.75-31.75
Tunica Media - Middle Coat: layer in the middle wall of the tubular structure	50.50-66.50
Nerves:	87.25-43.00
Valves: any one of various membranous structures in a hollow organ or passage that temporarily closes to permit flow of fluid in only one direction	71.50-57.20

Heart

The heart [0.00-70.00] is a hollow, muscular, contractile organ – the center of the circulatory system. It provides the propulsive force for circulating the blood throughout the vascular system. Its wall possesses three layers: the outer epicardium, a serous layer, the middle myocardium; and the inner endocardium, a layer that lines the four chambers of the heart and covers the valves. The heart is enclosed in a fibrous sac, the pericardium, the space between the pericardium and the epicardium forming the pericardial cavity.

Arteries: 36.50-23.60	
Left Coronary:	24.00-12.60
Interventricular Branch:	19.00-12.00
Right Coronary:	11.60-10.10
Interventricular Branch:	31.75-17.00
Marginal Branch:	13.75-10.50
Atria:	14.75-16.75
Septum:	11.25-20.50
Left:	22.75-18.80
Right:23.60-27.20	
Atrio-Ventricular Node:	21.25-16.40
Auricles: 14.20-14.60	
Ventricular Ring:	11.75-21.00
Bundle of His:	13.75-12.00
Endocardium:	11.50-11.10
Connective Tissue:	13.50-14.90
Elastic Fibers:	20.00-12.00

Fossa Ovalis:	18.90-16.50
Helpful Paired with Small Intestine [21.00-36.00]:	2.00-76.00
Pair with color red, emotion = happiness, sound = laughter	
Lymph Vessels:	22.00-27.40
Deep Plexus:	18.75-17.75
Superficial Plexus:	14.20-14.00
Myocardium:	20.00-19.00
Fibrous Muscular Rings:	10.20-22.75
Longitudinal:	10.75- 8.50
Transverse:	13.10- 4.00
Nerves:	11.25-20.75
Cardiac Plexus:	12.00-22.50
Deep:	22.50-17.20
Superficial:	14.50-15.00
Pericardium:	4.50-12.50
Arteries:	13.25-12.75
Fibrous Sac:	18.75-11.25
Lymph Glands, Mediastinal:	27.75-19.50
Nerves:	12.40-28.20
Serous Sac:	3.50-21.75
Purkinje's Fibers:	26.50-18.00
Purkinje's System:	18.20-22.60
Sino-Auricular Node:	12.00-19.50
Tendon of the Infundibulum:	30.75- 6.40
Valves:	16.00-14.00
Aortic:	8.00-10.60
Chordae Tendineae:	10.40-12.25
Mitral:	6.40-10.40
Papillary Muscles:	16.40-14.60
Pulmonary:	9.00- 5.60
Tricuspid:	11.50-10.00
Veins:	11.00- 8.50
Anterior Cardiac:	14.20-16.50
Coronary Sinus:	8.25-10.00
Valve of Coronary Sinus:	12.50- 8.30
Great Cardiac:	12.20- 9.30
Left Marginal:	13.60-15.30

Middle Cardiac:	7.75- 6.50
Oblique Vein of Left Atrium:	13.00-21.00
Posterior Vein of Left Ventricle:	4.75- 8.50
Right Marginal:	10.50- 8.25
Small Cardiac:	28.00-16.10
Ventricles:	10.75-16.25
Left:	13.50-16.50
Right:	6.60-11.30
Ventricular Septum:	17.60-18.75

Lymphatic System

The lymphatic system [35.00-37.50] is a network of delicate tubes throughout the body. It drains fluid (called lymph) that has leaked from the blood vessels into the tissues and empties it back into the bloodstream via the lymph nodes. The main roles of the lymphatic system is to manage the fluid levels in the body. *For detox, support and stimulate subtle fields of detox support program. Affected by Lyme Disease [44.00-41.00].*

Endothelium Lining to Glands:	38.75-43.60
Lymph Glands:	0.00-17.00, 3.00-scan, 36.00-17.00, 43.00-45.25
Auricularly: drains area of ear	32.25-38.75
Auxiliary: drains upper limbs and breasts	41.00-35.75
Bronchial: drains the lungs	37.20-41.20
Celiac: drains stomach, spleen, pancreas, liver	40.50-35.20
Deep Cervical: drains the tongue, lower pharnyx, larynx, and thyroid	39.25-29.50
Deep Inguinal: drains deep tissues of lower limbs *Use for arborvirus, arenavirus, bronchitis, emotional shock, emphysema, encephalitis, fright, parasites, and pulmonary disorders.*	45.00-29.00
Epitrochlear: drains the hands and forearms	37.75-37.00
External Iliac: drains the groin	35.00-28.80
Intercostal: drains the chest wall	36.25-27.20
Internal Iliac: drains the pelvic viscera	59.50-44.00
Lingual: drains deeper tissues of the neck	76.00-35.40
Lumbar: drains adrenals, kidneys, ovaries, testicles	67.00-38.50
Mediastinal: drains heart, pericardium, and thymus	48.80-29.80
Mesenteric: drains intestines	72.75-46.50
Popliteal: drains the legs	38.75-35.00
Sternal: drains front of chest wall and abdominal wall	47.00-35.50
Sub-Maxillary: drains face and floor of mouth	47.00-68.60
Sub-Occipital: drains back of the scalp	31.75-38.50
Superficial Cervical: drains external ear and neck	44.25-29.20

Lymph Vessels:

<u>External Coat:</u> 33.25-34.20

<u>Internal Coat:</u> 37.75-36.50

<u>Middle Coat:</u> 25.00-26.00

Lymph:

<u>Fluid:</u> 38.25-37.50

<u>Lymphocytes, Lymphoid Activity:</u> cloning sites for new immune cells, blood, red cells, white pulp 45.00-37.00

<u>Pericardial Lymph Fluid:</u> 39.50-43.00

<u>Vessels:</u> 34.75-27.60

Right Lymphatic Duct, Tributaries:

<u>Right Broncho Mediastinal Trunk:</u> 33.75-27.25

<u>Right Jugular Trunk:</u> 30.50-36.00

<u>Right Subclavian Trunk:</u> 24.00-38.25

Thoracic Duct: 26.00-33.00

<u>Cisterna Chyli:</u> 29.75-29.50

<u>Tributaries:</u>

<u>Left Broncho Mediastinal Trunk:</u> 38.50-37.00

<u>Left Jugular Trunk:</u> 31.25-25.50

<u>Left Subclavian Trunk:</u> 38.25-37.75

Spleen: 0.00-95.00, 4.00-92.50, 88.50- 9.50

The spleen is a small organ inside your left rib cage, just above the stomach. It's part of the lymphatic system (which is part of the immune system). The spleen stores and filters blood and makes white blood cells that protect you from infection. Many diseases and conditions can affect how the spleen works.

<u>Arteries:</u>

<u>Arterioles:</u> 39.75-45.25

<u>Branches in Spleen:</u> 37.00-55.00

<u>Splenic:</u> 36.00-66.50

<u>External Serous Coat:</u> 13.25-17.50

<u>Helpful Paired with Stomach [77.00-84.00]:</u> 4.00-95.00
Pair with color yellow, emotion = pensiveness, sound = singing

<u>Internal Fibro-elastic Coat:</u> 22.25-26.50

<u>Splenic Cells:</u> 32.25-31.50

<u>Splenic Pulp:</u> 38.60-34.50

<u>Trabeculae:</u> 41.75-26.00

<u>Tunica Albuginea:</u> the white fibrous coating of the eye, testicle, ovary, or spleen 26.75-20.25

Lymph Vessels:	58.50-38.90
Lymphatic Nodules:	62.25-47.20
Nerves:	44.00-35.00
From Coeliac Plexus:	37.25-42.50
Peritoneum:	33.75-66.25, 91.00-31.00
Veins:36.50-43.10	
Tributaries:	31.25-29.25

Digestive System

The digestive system breaks down food into nutrients so they can then be absorbed into the bloodstream and used by the body for energy, growth and repair.

Mouth: 11.00-81.00

The digestive process starts in your mouth when you chew. Your salivary glands make saliva, a digestive juice, which moistens food so it moves more easily through your esophagus into your stomach. Saliva also has an enzyme that begins to break down starches in your food.

Arteries: 21.25-19.50

Gums: 3.00-25.00, 51.00-25.00

Lip, Bottom: 32.75-34.75

Lip, Top: 31.50-34.50

Lymph Vessels: 76.40-56.50

Nerves: 41.00-42.00

Pharnyx: 21.75-33.00, 24.00- 7.00

- Adenoids: 33.50-46.50, 76.00-36.00
- Constrictor Muscles: 26.25-23.75
- Epiglottis: thin, triangular, lid-like flap of cartilage that folds over the opening of the windpipe during swallowing to prevent foreign matter from entering the lungs 0.00-16.00, 0.00-42.50, 15.50-43.50
- Eustachian Tubes: 69.50-38.00
- Mucous Coat: 57.40-33.00
- Nerves: 33.00-40.75
  - Glossopharyngeal Branch: 75.00-16.00, 0.00-26.50
  - Of Sphenopalatine Ganglion: 29.75-24.00
- Stratified Epithelium: 38.00-37.90
- Tonsils:
  - Lymphocytes: 16.75-33.25
  - Lymphoid Tissue: 17.50-28.00
  - Tonsillar Artery: 25.40-15.75
  - Tonsillar Plexus: 26.25-24.75
  - Uvula Palatina: 29.75-24.25

Salivary Glands: 30.75-36.25

- Parotid Gland:
  - Arteries: 39.00-12.50
  - Jaws: 39.00-24.00

<u>Lymph Vessels:</u>	10.50-18.50
<u>Nerves:</u>	56.60-52.00
• Auriculotemporal Branch:	2.20-49.40
• Of Otic Ganglion:	77.75-66.50
<u>Veins:</u>	22.20-30.25
Sublingual Gland:	16.00-23.50
Submaxillary Gland:	22.75-14.50
Teeth:	4.00-99.00, 59.00-20.00
Tongue:	85.00-43.00
<u>Circumvallate Papillae:</u>	0.40-24.50
<u>Extrinsic Muscles:</u>	11.75-46.25
<u>Filliform Papillae:</u>	41.50-77.00
<u>Fungiform Papillae:</u>	10.50- 7.80
<u>Mucous Membrane:</u>	12.50-59.00
<u>Pharynx, Eustachian Tubes, Taste Buds:</u>	0.00-43.50
<u>Bitterness:</u>	30.00-50.25
<u>Sourness:</u>	32.25-36.50
<u>Sweetness:</u>	15.50-29.75
<u>Veins:</u>42.00-49.25	

Esophagus: A hollow, muscular tube that moves food and liquid from your mouth to your stomach.	17.25-29.25, 20.00-43.00
Mucous Membrane: the membrane lining passages and cavities communicating with the air, consisting of a surface layer of epithelium, a basement layer, and an underlying layer of connective tissue – the lamina propria	19.40-21.25
Muscle Fibers:	18.75-24.50
<u>Muscle Fibers, Circular:</u>	20.00-24.00
<u>Muscle Fibers, Longitudinal:</u>	13.50-19.00
Stratified Epithelium:	20.00-21.00
Sub-Mucous Coat:	23.75-24.00

Stomach:

The stomach [0.00-32.00] is a muscular organ located on the left side of the upper abdomen. The stomach receives food from the esophagus. The stomach secretes acid and enzymes that digest food.

Arteries:	25.00-31.00
Gastric, Left:	42.50-39.00
Gastric, Right:	29.90-34.00
Gastroepiloic, Left:	21.50-22.75
Gastroepiloic, Right:	31.00-42.60
Gastric Glands:	23.75-23.25
Cardiac:	37.75-36.20
Fundus: source of Hydrochloric Acid (HCl)	19.20-30.00
Pyloric: source of pepsin	29.00-23.80
Gastric Lymphatic Nodules:	31.25-33.60
Helpful Paired with Spleen [4.00-95.00]: *Pair with color yellow, emotion = pensiveness, sound = singing*	77.00-84.00
Inner Mucous Coat:	24.25-22.75
Columnar Epithelium:	39.00-42.50
Gland Ducts:	45.25-49.00
Goblet Cells: a cell found in the epithelium of the intestinal and respiratory tracks that secrets mucous	47.25-32.50
Peptic Glands:	45.00-39.75
Large Intestine, Omentum:	0.00-52.50
Lymph Vessels:	17.50-22.50
Muscular Coat:	31.75-29.75
Circular Fibers:	45.25-43.10
Longitudinal Fibers:	39.60-31.50
Oblique Fibers:	41.75-44.50
Nerves:	27.75-21.00
Parasympathetics:	
From Vagus Nerve:	41.75-47.10
Greater Anterior Gastric Nerve:	42.00-48.20
Greater Posterior Gastric Nerve:	40.75-47.50
Plexuses in Muscular Coat:	40.00-44.50
Plexuses in Submucous Coat:	46.00-37.50
Sympathetics:	
From Cellac Plexus:	33.90-51.60
From Left Phrenic Plexus:	44.00-62.50
From Left Splanchnic Nerve:	46.75-55.75

From Lumbar Sympathetic Trunk: 31.25-37.50

From Plexuses Around Epiploic Artery: 34.00-58.20

From Plexuses Around Gastric Artery: 35.75-41.30

From Thoracic Sympathetic Trunk: 48.60-53.50

Omentum: a double fold of peritoneum attached to the stomach and connecting it with certain of the abdominal viscera 31.25-22.00

Pylorus: the lower orifice of the stomach, opening into the duodenum 0.00-38.00, 23.75-24.50

Serous Coat: 47.50-28.25

Sub Mucous Coat: 23.75-29.75

Blood Vessels: 43.20-34.50

Lymphatics: 19.75-36.50

Veins: 26.00-32.00

Between Glands: 39.20-46.25

Mesenteric, Superior: 36.20-42.40

Portal: 39.50-59.00

Splenic: 35.50-38.75

Duodenum

The Duodenum [34.25-43.00] is the first part of the small intestine connecting with the pylorus of the stomach and extending to the jejunum. *Balance 16.25-25.00 if DDT, insect stings, migraines, parathion, poisons, acidity in burns, or lindane is found.*

Lymph Vessels: thin-walled vessels conveying lymph from the tissues, resembling veins in structure, possessing three layers: the intima, media, and adventitia; and having paired valves. 22.75-21.25

Mucous Membrane: the membrane lining passages and cavities communicating with the air, consisting of a surface layer of epithelium, a basement membrane, and an underlying layer of connective tissue – the lamina 26.25-42.25

Circular Folds: 17.20-28.00

Duodenal Glands: 28.25-41.00

Villi: the short, filamentous processes found on some membrane surfaces 22.25-17.00

Artery Leaving Villus: 24.50-26.25

Capillaries: 27.00-25.00

Columnar Epithelium: pillar-shaped cells 11.10-12.50

Goblet Cells: secretory cells found in the epithelium of the intestinal and respiratory tracts; a unicellular gland that secretes mucus. 22.25-25.10

Lacteals: an intestinal lymphatic that takes up chyle and passes it to the the lymph circulation and the blood vascular system 30.25-30.50

Lymph Vessels: thin-walled vessels conveying lymph from the tissue 36.60-27.00

Muscle Cells 25.00-27.00

Muscular Coat: the layer of muscles on a given structure	43.25-22.50
Circular Fibers:	39.50-50.25
Longitudinal Fibers:	63.50-60.20
Nerves:	
From Splanchnic Nerves: three nerves connecting the thoracic sympathetic ganglia distributed to the viscera	35.00-21.75
From Vagus Nerve:	41.25-36.60
Meissner's Plexus, Vagal Origin: small aggregations of ganglion cells located in submucosa of intestine	29.75-33.75
Myenteric Plexus, Vagal Origin: involving the muscular coat of the intestine	25.60-23.60
Sympathetics, Inhibit Peristalsis:	22.25-21.75
Peritoneal Covering: the serous membrane	23.00-21.75
Sub-Mucous Coat: beneath the mucous membrane	34.75-48.75
Blood Vessels:	36.50-55.50
Lymphatics:	38.10-55.25
Nerves:	35.50-37.40
Suspensory Muscle: supporting structure in the body	34.25-52.25
Veins:	31.50-31.25

Small Intestine

The small intestine [0.00-82.00] helps to further digest food coming from the stomach. It absorbs nutrients (vitamins, minerals, carbohydrates, fats, proteins) and water from food so they can be used by the body.

Arteries:	
Ileal:	74.00-63.00
Jejunal:	26.60-34.10
Mesenteric Superior:	26.40-45.90
Helpful Paired with Heart [2.00-76.00]: *Pair with color red, emotion = happiness, sound = laughter*	21.00-36.00
Ileum: lowest part, opening to large intestine	20.50-22.00
Lymph Vessels:	22.50-35.50
Mucous Membrane:	51.75-42.25
Aggregated Lymphatic Folds: Peyer's patch	2.00-22.00
Circular Folds:	61.75-36.20
Solitary Lymphatic Nodules:	11.75-31.50
Villi:	42.10-59.10
Artery Leaving Villus:	81.50-17.60
Capillaries:	17.25-11.75
Columnar Epithelium:	22.75-25.25
Goblet Cells:	26.25-27.75

<u>Lacteals:</u>	3.80-16.50
<u>Lymph Vessels:</u>	23.00-28.50
<u>Muscle Cells:</u>	72.50-66.00
Muscular Coat:	13.75-26.50
<u>Circular Fibers:</u>	35.50-30.00
<u>Longitudinal Fibers:</u>	16.25-24.00
Nerves From:	34.25-44.50
<u>Coeliac Ganglion:</u>	60.75-70.50
<u>Meissner's Plexus:</u> from which fibers also pass to mucous membrane	49.50-49.25
<u>Myenteric Plexus:</u>	46.10-56.60
<u>Plexus Around Superior Mesenteric Artery:</u>	59.00-59.00
<u>Splanchnic Nerve:</u>	46.75-58.00
<u>Sympathetics, Inhibit Peristalsis:</u>	45.00-39.25
<u>Vagus Nerve:</u>	71.40-72.60
Parasympathetics: Augment Peristalsis and secretion of gastric juice	56.60-43.00
Plexus of Minute Branches:	73.50-72.00
Serous Coat:	18.00-19.50
Sub Mucous Coat:	26.75-53.50
<u>Blood Vessels:</u> including arteries, capillaries, and veins	9.50-28.00
<u>Lymphatics:</u>	32.50-37.10
<u>Nerves:</u>	37.75-38.50
Veins:	35.50-44.50

Large Intestine

The long, tube-like organ that is connected to the small intestine at one end and the anus at the other. The large intestine has four parts: cecum, colon, rectum, and anal canal. Partly digested food moves through the cecum into the colon, where water and some nutrients and electrolytes are removed.

Anus:	0.00-42.50, 28.25-24.00
Ascending Colon:	0.00- 8.00, 35.00-47.25
Cecum: 15.00-35.00, 90.00-66.00	
<u>Transverse Colon:</u>	0.00-60.00
Descending Colon:	0.00-62.00, 0.00-90.00, 16.40-15.00, 23.75-14.00
Helpful Paired with Lungs [92.00-83.00] *Pair with color white, emotion = sadness, sound = weeping*	8.00-72.00
Hepatic Flexure:	0.00-78.00
Ileocecal Valve:	21.00-69.00
<u>Frenula:</u>	10.50-39.50

Lymph Vessels: 75.50-80.25

Mesentery: 0.00- 0.50, 33.75-43.50

Mucous Membrane: 11.25-14.75

Muscular Coat: 21.75-28.00

Circular Fibers: 16.40-18.00

Longitudinal Fibers: 20.00-12.00

Omentum: 29.75-44.00

Pelvic Colon: 0.00-33.75, 50.75-22.75

Pelvic Mesocolon: 28.25-20.50

Rectum: 8.50-90.00, 12.00-35.00, 50.00-12.00, 85.00- 1.00

Also balance with 55.00-96.00, then 40.00-96.00

External Sphincter: 90.25-87.00

Internal Sphincter: 77.25-75.25

Perineal Flexure: 77.50-77.00

Rectal Ampulla: 90.50-87.50

Sacral Flexure: 72.00-67.00

Serous Coat: 31.00-18.50

Sigmoid Flexure: 0.00-12.00, 28.75-42.50

Solitary Lymphatic Nodules: 18.00-17.50

Structure of Arteries:

Mesenteric Inferior: to rectum and descending, transverse, and pelvic colons 85.00-66.00

Rectal Inferior: 32.75-29.25

Rectal Middle: 25.75-26.50

Rectal Superior: 25.50-31.00

Mesenteric Superior: to cecum, appendix, ascending and transverse colons 64.75-57.50

Transverse Colon: 41.75-39.25

Left Colic Flexure: 26.20-10.40

Phrenicocolic Ligament: 13.00- 6.00

Transverse Mecocolon: 17.90-16.20

Veins:

Hemorrhoidal Venous Plexus: 83.75-56.00

Nerves: Parasympathetic & Pelvic Splanchnic: 79.50-68.75

Vagus Branch of Large Intestine: 80.25-62.00

Nerves: Sympathetics:

Celiac Ganglion: 70.50-77.75

Hypogastric Plexus: 77.25-62.75

Lumbar Plexus: 56.00-48.25

Superior Mesenteric Ganglion: 49.75-59.50

Vermiform Appendix: 0.00-66.00, 3.00- 8.00, 40.00-46.00

Use for eczema, flatulence, malaria, thyroid, typhoid fever, viral and respiratory viral diseases.

- Circular Muscular Fibers: 13.50- 7.20
- Longitudinal Fibers: 23.60-44.60
- Membrane: 17.25-20.75
- Serous Coat: 44.50-41.00
- Submucous Coat: beneath the mucous membrane 23.60-19.20

Gallbladder & Biliary Ducts

The gallbladder is a pear-shaped sac on the undersurface of the right lobe of the liver that holds bile from the liver until discharged through the cystic duct. In addition, the gallbladder acts to concentrate the bile by removing its water content.

Biliary Ducts, Bile Duct: 0.00-23.50, 50.50-52.75

- Cystic Duct: 41.50-51.50
- Common Hepatic Duct: 57.00-44.50
- Spiral Valve: 50.75-47.50

Gall Bladder: 0.00-35.00, 14.00-64.00

Balance with 13.00-64.00.

- Arteries: 36.50-27.50
  - Cystic: 48.25-45.50
- Fibromuscular Coat: 27.00-37.75
- Helpful Paired with Liver [17.00-29.00]: 13.00-64.00
  Used for constipation, lindane, colds, intestinal worms, anger/shouting. Pair with green.
- Lymph Vessels: 51.50-32.75
- Mucous Coat: 24.40-23.80
- Nerves: 44.50-47.50
  - Celiac Branches: 24.00-27.20
  - of Hepatic Plexus: 18.25-27.50
  - Right Phrenic Fibers: 20.00-12.00
- Serous Coat: 23.75-14.00
- Sphincter: 30.25-27.00
- Veins:29.25-38.25
- Veins, Cystic: 80.00-37.00

Liver

The liver [0.00-29.00] is the largest solid organ in the body. It removes toxins from the body's blood supply, maintains healthy blood sugar levels, regulates blood clotting, and performs hundreds of other vital functions. It is located beneath the rib cage in the right upper abdomen.

Bile Pigment: 42.50-40.25

Filter Factor: 7.50-29.00

Hepatic Artery: 45.50-40.75

- Conveying Arterial Blood To:
  - Bile Ductules: 29.00-33.00
  - Connective Tissue: 33.00-46.10
  - Interlobular Plexuses: 7.25-20.50
  - Walls of Portal Vein: 33.50-34.00
- Conveying Oxygenated Blood To:
  - Bile Capillaries: 25.50-29.00
  - Bile Ductules: 33.25-30.25
  - Liver Cells: 21.25-35.50, 41.00-39.75
  - Sinusoids: 30.75-36.25

Left Lobe: 0.00-45.00, 43.20-28.00, 45.00-29.00

Check for allergies, cancer, constipation, diabetes, edema, insect stings, leukemia, mumps, phlegm, poisons, protein digestion, and undulant fever.

- Coronary Ligament: 29.00-38.00
- Falciform Ligament: 29.00-18.00
- Left Triangular Ligament: 36.00-28.00
- Right Triangular Ligament: 42.50-48.25

Liver Factor: 12.00-29.00

Lymph Vessels: 45.00-29.25

- Deep: 40.00-38.50
- Superficial: 51.75-49.50

Nerves: 52.00-36.00

- From Celiac Plexus: 44.25-39.00
- From Right Phrenic: 33.00-16.00
- From Vagus Nerves: 30.75-34.50
- Sympathetics: 45.50-25.00

Right Lobe: 18.00-29.00, 27.75-30.00, 72.00-29.00

- Caudate Lobe: 47.50-37.00
- Quadrate Lobe: 40.25-35.00
- Spigelius's Lobe: 36.75-48.50

Surfaces:	
Anterior Surface:	30.75-43.25
Fissure for Ligamentum Venosum:	40.30-39.00
Inferior of Visceral Surface:	40.25-20.75
Colic Impression:	27.00-27.50
Duodenal Impression:	34.75-24.50
Gastric Impression:	35.00-26.75
Porta Impression:	26.75-24.00
Renal Impression:	35.00-31.50
Posterior Surface:	34.00-38.75
Groove for Inferior Vena Cava:	46.75-41.50
Suprarenal Impression:	54.25-38.50
Right Surface:	36.75-52.00
Superior Surface:	34.25-40.75
Veins:	20.00-23.75
Hepatic:	31.00-41.50
Portal:	30.20-36.25

Pancreas

The pancreas is a compound tubuloacinar or racemose gland situated behind the stomach in front of the first and second lumbar vertebrae in the horizontal position, its head attached to the duodenum and its tail reaching into the spleen. The gland is composed of lobules that form lobes connected by strands of tissue with ducts that lead from the lobules into the Duct of Wirsung, the main pancreatic duct.

The pancreas produces both internal and external secretions. The internal secretions, which are elaborated by the Islets of Langerhans, include the hormones insulin and glucagon. In conjunction with hormones from other endocrine glands, these play a role in the regulation of carbohydrate metabolism. The external secretion, called pancreatic juice, is produced by the cells in the acini. It passes through the pancreatic ducts into the duodenum, where it plays a role in the digestion of all foods.

Pancreas: stimulates enzyme production	0.00-34.00, 9.00-70.00, 85.50-22.50
Chromium needs to be in balance for proper function.	
Arteries, from Branch of Hepatic Artery:	25.50-23.50
Arteries, from Branch of Mesenteric Artery:	29.75-23.25
Arteries, from Splenic Artery:	20.25-39.50
Head of Pancreas:	42.25-43.50
Anterior Surface:	23.50-26.00
Posterior Surface:	29.60-32.25
Insulin-Free Section:	43.00-53.00

Lymph Vessels:	24.25-37.50
Nerves	
From Splanchnic Nerves: three nerves from the thoracic sympathetic ganglia distributed to the viscera	27.25-25.25
From Vagus Nerve:	20.25-26.00
Through Splenic Plexus:	29.60-32.50
Pancreatic Duct:	27.50-21.75
Accessory of Pancreatic Duct:	27.25-35.00
Ampulla of Bile Duct:	27.75-31.75
Structure:	
Alveoli:	36.00-35.25
Centro-acinar Cells:	35.75-33.00
Intercalcary Ducts:	6.00- 9.25
Islets of Langerhans:	8.00-91.00, 36.00-35.00
• Alpha Cells: Produces glucagon, balances blood sugar	32.00-24.00
• Beta Cells:	25.00-26.00
Paranucleus:	33.50-21.75
Sugar-Free Part, Protein:	6.00- 4.00
Tail of Pancreas:	21.75-30.25
Anterior Surface:	27.50-30.00
Inferior Surface:	25.20-28.75
Posterior Surface:	32.80-22.50
Superior Border:	24.25-24.00
Veins:	22.00-31.75

Musculoskeletal System

Muscular Structure

Aponeuroses:	91.50-68.00
Fasciae/Fascia Tissue: surrounds organs and muscles, helping to keep their shape. Emotions may settle into these tissues, locking up the fascia.	73.50-81.50
Deep Fascia:	64.50-57.10
Fatty Nodules:	79.20-73.50
Superficial Fascia:	79.60-69.25
Synovial Bursae:	73.00-87.25
Synovial Membrane:	76.25-73.50
Synovial Sheaths:	68.50-61.30
Ligaments:	7.00-14.50, 21.50-45.50, 58.75-65.00
Muscular Tissue:	0.00-41.00, 46.00-36.50
Muscular Tissue, Tics and Twitches:	11.00-41.00
Blood Supply:	58.25-43.50
Cardiac:	86.50-48.40
Muscle Cells:	58.50-51.00
Sarcolemma:	57.25-22.25
Muscle Sheath:	53.50-53.25
Sphincter:	0.00-42.50
Nerve Supply:	39.75-49.00
Motor Nerves:	37.75-58.50
• End Plates:	26.50-28.00
With depression, check the need to increase or decrease acetylcholine.	
Sensory Nerves:	90.25-81.00
• Muscle Spindles:	66.60-63.00
Striped or Voluntary:	61.50-75.00
Tendons, Ligaments:	0.00-45.50
Unstriped or Involuntary:	0.00-57.00, 56.50-61.00
Tendons:	57.50-69.00
Nerve Supply:	75.60-69.50
Sheaths:	55.00-64.50

Skeletal Structure

Bone:	25.00-22.00, 41.50-42.60
Blood Vessels: including arteries, capillaries, and veins	27.25-36.40
Canaliculi:	90.75-82.40
Haversian Canals:	38.25-40.00
Lacunae:	78.50-80.25
Nutrient Artery:	67.00-61.00
Marrow: (also see Reticuloendothelial System)	1.00-4.40, 35.00-92.00, 0.00-21.00
Nerves:	83.00-68.00
Osteogenetic Tissue: young bones only	39.60-44.10
Periosteum:	35.40-30.00
Cartilage:	33.00-36.00
Cellular Cartilage:	57.75-46.50
Elastic Fibrocartilage:	48.50-51.25
Hyaline Cartilage:	39.50-42.25
White Cartilage:	39.75-42.00
Joints:	79.00-52.00
Vertebrae:	34.75-34.25
Intervertebral Discs:	38.75-40.25

Reproductive Systems

Female

Fallopian Tubes 53.00-50.50

- External Serous Coat: 31.75-27.50
- Fimbriated End: 47.50-32.25
- Infundibulum: 26.50-33.50
- Internal Mucous Coat: internal layer on a tubular structure 50.50-33.50
- Middle Muscular Coat: 47.50-48.00
- Tube, Left: 23.00-19.00
- Tube, Right: 23.00-17.00

Ovaries: 0.00-22.00, 0.00-59.50, 3.00-54.00, 82.00-65.50

- Corpus Luteum: 58.25-45.75
- Germinal Epithelium: 44.10-44.30
- Graafian Follicles: a mature vesicular follicle of the ovary 48.25-46.00
- Membrane Granulosa: layers of cells in the graafian follicle 47.00-49.00
- Ovum: 41.00-46.50
- Stroma: foundational, supporting tissues of an organ 46.75-23.00
- Suspensory Ligament: 11.60-35.30
- Tubal Extremity: pertaining to the fallopian tube 1.50-56.25
- Tunica Albuginea: white fibrous coating of the ovary 26.75-20.25
- Uterine Extremity: 15.50-20.00

Uterus: 0.00-58.00, 8.00-23.00, 10.00-61.00

- Arteries, from Internal Iliac Artery: 61.50-51.00
- Body of Uterus: 42.75-40.50
  - Fundus: portion of an organ most remote from its opening 33.00-40.25
- Cervix of Uterus: 0.00-26.00, 27.40-37.40
  - External OS: 46.75-43.50
  - Internal OS: 40.25-69.75
  - Mucous Glands: 82.50-70.00
  - Parametrium: loose connective tissue around the uterus 38.75-43.20
- External Serous Coat: 0.00-53.50, 50.00-32.60
- Internal Mucous Coat: internal layer on a tubular structure 47.75-39.50
- Ligaments:
  - Anterior Ligament: 8.00- 5.50
  - Posterior Ligament: 20.00-17.50
  - Round Ligaments: attached below and in front of the fallopian tube 87.25-64.75

Two Broad Ligaments: attached to lateral borders of uterus from insertion of the fallopian tube to above the pelvic wall	15.00-62.00
Lymph Vessels:	86.25-10.50
Muscular Coat:	39.20-18.25
Nerves:	
From Hypogastric Plexus:	41.25-19.00
From Ovarian Plexus:	19.00-18.50
Pelvic Visceromotor Fibers: conveys motor impulses to the viscera	39.40-28.00
Veins, Ending in Uterine Plexuses:	62.50- 8.25
Vagina:	92.00-31.00, 92.00-96.00
Anterior Wall:	28.10- 6.00
Arteries:	51.25-60.25
Clitoris:	77.10-86.10
Erectile Tissue:	38.25-17.50
Labium:	60.60-36.25
Lymph Vessels:	60.75-11.25
Mucous Membrane:	11.10-10.00
Muscular Coat:	46.50-39.00
Nerves:	
From Pelvic Visceromotor Fibers:	39.75-89.10
From Vaginal Plexuses:	26.75-15.00
Perineal Body:	63.75-46.75
Posterior Fornix: fibrous, vaulted band connecting to the uteri	31.00-16.25
Posterior Wall:	24.50-15.00
Veins:	49.10-35.25
Vulva:	0.00-51.50, 15.00-18.20

Male

Penis:	0.00-76.50, 6.00-88.00
Arteries:	64.25-58.00
Helicine Arteries:	77.50-64.00
Body:	64.00-54.25
Corpus Spongiosum Penis:	82.00-71.50
Glans Penis:	75.50-89.00
Neck of Penis:	88.75-46.50
Bulb of the Penis:	35.25-52.25

Nerves: 68.25-59.00

From 2nd Sacral: 44.75-52.20

From 3rd Sacral: 75.50-86.50

From 4th Sacral: 52.50-52.75

Root: 71.25-54.25

Crus: 33.60-45.00

Veins: 82.00-69.50

Prostate Gland: 0.00-26.00, 5.00-91.00, 13.00-26.00, 13.50-26.00, 52.75-82.50, 58.00-26.00

Surrounds the neck of the bladder and urethra in males; partly glandular and part muscular. It secretes a thin, opalescent, slightly alkaline fluid that is part of the semen

Arteries: 57.50-51.75

Glandular Substance: 73.75-63.75

Glandular Tissue: 85.00-56.00

Interstitial Tissue: 83.25-61.50

Lateral Lobes: 84.25-61.00

Lymph Vessels: 71.75-57.75

Muscular Tissue: 62.75-69.00

Nerves: 19.50-95.00

Veins:13.75-21.50

Scrotum: 2.00-98.00, 25.60-25.75

Dartos Muscle: 36.60-33.00

Ejaculatory Ducts: 81.50-48.60

Muscular Coat: 68.25-53.50

Nerves, Derived from Pelvic Plexuses: 40.25-37.00

Outer Fibrous Layer: 22.25-48.50

Seminal Vesicles: 75.50-63.50

Skin: 27.50-35.50

Spermatic Cord: 38.10-36.50

Arteries: 56.50-50.25

Fascia, Cremasteric: 30.75-47.60

Fascia, External Spermatic: 57.25-55.00

Fascia, Internal Spermatic: 31.75-34.75

Nerves: 54.25-44.50

Testes:	0.00-22.00, 37.00-94.00, 37.25-35.50
Epididymis:	34.40-33.50
Ductule Aberrant: one of a group of tubules associated with the epididymis	77.25-48.75
Lymph Vessels:	32.75-31.50
Nerves:	27.50-27.60
From 10th Thoracic Segment:	73.00-46.50
From Aortic Segment:	68.25-57.50
From Renal Plexus:	72.25-30.50
Testicular Artery:	33.25-26.75
Testicular Veins:	26.75-12.75
Pampiniform Plexus:	29.50-27.25
Tunica Albuginea: the white, fibrous coat of the testicle	26.75-20.25
Tunica Vaginalis: serous membrane surrounding the front and sides	30.25-28.00
Tunica Vasculosa:	28.50-18.50
Vas Deferens: the excretory duct of the testes; the continuation of the canal of the epididymis	36.60-59.75

Respiratory System

The organs involved in the interchange of gases between an organism and the atmosphere, consisting of the bronchi, larynx, lungs, nose, pharynx, and trachea.

Nose:	0.00-23.50, 19.00-96.00, 51.00-53.00
Anterior Nares:	14.75-21.25
Lower Nasal Cartilage:	21.00-53.00
Nasal Cavity:	20.75-17.25, 32.00-48.00
Nerves, Facial:	45.75-51.00
Nerves, Skin:	40.75-50.60
Olfactory Region:	16.25-17.00
Posterior Nares:	17.00-16.25
Respiratory Region:	20.00-20.75
Arteries:	22.60-23.50
Lateral Wall:	20.50-27.25
Nasal Mucous Membrane:	20.60-22.75
Nasal Septum:	23.40-20.00
Nasolacrimal Duct:	20.75-19.50
Nerves:	11.50-12.50
Veins:	24.00-21.50
Upper Nasal Cartilage:	9.75-16.50
Bronchi:	0.00-83.00, 4.00-74.00, 92.00-82.00
Arteries:	42.00-56.00
Bronchioles:	63.00-74.00
Cartlidge Rings:	47.25-65.20
Lymph Vessels:	32.00-20.75
Passing to Paratracheal Gland:	29.00-18.25
Passing to Pretracheal Gland:	22.75-29.25
Mucous Membrane:	46.00-46.00
Muscular Tissue:	62.00-39.50
Nerves:	29.75-28.50
From Recurrent Laryngeal Nerves:	39.75-31.25
From Sympathetic Trunks:	35.75-29.00
From Vagus Nerve:	37.00- 7.75
Smooth Muscle:	51.75-42.50
Sub-Mucous Layer:	40.75-24.50
Veins:	50.40-42.25

Diaphragm, Muscles: 37.25-71.50

Larynx: the "voicebox" 0.00-16.00, 32.00-16.00, 61.50-47.75

- Arteries: 57.00-39.00
- Ligaments: 32.25-39.00
- Mucous Membrane: 66.00-49.50
- Muscles: 38.50-36.20
- Nerves: 40.25-43.30
- Nine Cartilages: the laryngeal skeleton made up of the thyroid cartilage, cricoid cartilage, epiglottis, arytenoid cartilages, corniculate cartilages, and cuneiform cartilages. 59.50-33.25
- Veins:50.00-43.25

Lungs: 0.00-47.00, 4.00-83.00, 88.25- 8.60

May be helpful to pair with Large Intestine (8.00-72.00). Affected by Lyme Disease. Supported by neutralizing toxins and tobacco. Pair with color white, emotion = sadness, sound = weeping.

- Arteries: 31.00-91.00, 31.00-93.00
  - Pulmonary Artery: 50.75-64.75
- Intercostal Muscles, External: 52.50-39.50
- Intercostal Muscles, Internal: 52.50-42.50
- Intrapulmonary Bronchi: 85.75-63.50
  - Alveoli: 28.00-14.75
  - Bronchial Muscles: 50.00-35.00
  - Mucous Membrane: 34.50-34.25
  - Outer Fibrous Coating: 62.25-58.75
  - Terminal Bronchioles: 38.50-13.50
- Nerves: 37.50-76.50
  - Broncho-constrictors from Vagus Nerve: 39.25-46.00
  - From Vagus Nerve: 57.60-69.50
  - Pulmonary Plexus Anterior: 49.00-39.50
  - Pulmonary Plexus Posterior: 60.00-44.00
- Pulmonary Substance: 39.50-30.75
  - Lobules: 59.00-42.50
- Serous Coat: 74.75-28.50
- Sub-serous Tissue: 54.00-40.00, 54.00-50.00
- Thoracic Area, Thorax, Chest: 0.00-53.00
- Veins: 41.00-53.75

Mediastinum: a septum or cavity between two principal portions of an organ 40.50-62.50

Pleurae: 53.25-51.30

- Arteries: 26.75-43.00
- Basement Membrane: 53.25-72.00
- Lymph Vessels: 95.25-37.60
- Nerves: 79.00-69.25
  - From Phrenic Nerve: 42.75-33.50
- Pleural Cavity: 57.25-62.25
- Serous Coat: 46.25-44.50
- Unstriped Muscle: 34.50-34.50

Trachea: 31.75-75.40

- Cartilages: 73.25-69.00
- Mucous Membrane: 56.25-49.75

Sinuses

The sinuses [0.00-42.75] are air-filled spaces in the skull. They are located behind the forehead, nasal bones, cheeks, and eyes. Healthy sinuses contain no bacteria or other germs. Most of the time, mucus can drain out and air is able to flow through the sinuses.

Air Sinuses: 5.50-33.00, 26.00-33.00

Balance 10.00-20.00 to clear.

- Antrum: 36.40-31.25
- Ethmoid: 28.40-47.00
- Frontal: 42.75-39.00
- Mastoid: 41.50-37.50
- Maxillary: 0.00-23.25
  - Lower Lungs: 32.00-33.00
- Sphenoid: 27.75-18.75
- Sphenoidal, Paranasal Sinuses: 0.00-37.50

Paranasal Sinuses: 32.75-27.70

- Antrum: 44.50-45.50
- Ethmoidal: 33.90-43.00
- Frontal: 0.00-3.00, 0.00-50.00, 36.75-37.00
- Maxillary: 0.00-33.00, 32.50-20.00
- Nasal Mucous Membrane: 43.00-46.00
- Sphenoidal: 31.50-37.00

Venous Sinuses: 11.25-17.50

Coronary: 11.25-17.50

Dura Mater, Antero-Inferior Group:

Basilar Network: 46.50-47.20

Cavernous: 43.25-47.20

Intercavernous: 44.20-48.40

Middle Meningeal: 50.60-58.00

Petrosal Inferiour: 49.50-49.00

Petrosal Superior: 57.50-53.00

Dura Mater, Draining Blood from Brain: 21.25-15.50

Dura Mater, Postero-Superior Group:

Occipital: 51.00-56.00

Sagittal Inferior: 58.40-50.20

Sagittal Superior: 54.25-58.00

Sigmoid: 45.90-40.25

Straight: 46.50-53.00

Transverse: 43.00-46.50

Reticuloendothelial System

The Reticuloendothelial System is applied to those cells scattered throughout the body that have the power to ingest particulate matter. Includes macrophages of loose connective tissue; reticular cells of lymphatic organs and myeloid tissues; Kupffer cells of the liver; cells lining blood sinuses of spleen, bone marrow, adrenal cortex, and hypophysis; microglia of the central nervous system; adventitial cells of blood vessels; and dust cells of the lungs.

Cells Comprising Loose Reticulum In:

Connective Tissues:	28.00-30.20
Endotheliocyte, Classmatocytes:	42.50-50.25
Endothelium Lining the Vascular Sinuses:	10.75-26.00
Anterior Pituitary:	35.25-35.00
In the Bone Marrow:	11.75- 9.50
Liver, Kupffer's Cells:	29.00-30.00
Lymph Glands:	17.00-20.00
Medulla of Suprarenal Gland:	18.20-18.40
Spleen:	14.00-20.75
Meninges:	30.40-59.00
Meningocytes:	3.50- 5.50
Microglia Cells:	
Blood:	45.50-50.25
Mononuclear Leucocytes:	36.75-18.50
Neuroglia:	10.00- 6.30

Cells of Bone Marrow:

Red Marrow:	25.10-18.50
Erythroblast:	41.20-56.00
Erythrocytes:	34.00-26.50
Myelocytes:	41.75-39.50
Normoblast:	29.50-52.25
Reticulocyte:	57.00-64.00
Reticulo-Endothelium:	42.75-39.00
Megakaryoblasts:	83.25-65.25

<u>Megakaryocytes:</u> evolves into thrombocytes or blood platelets at their death, giving off the chemical serotonin, which carries messages between nerve cells in the brain and throughout your body	93.60-85.75
<u>Plasma Cells, Activated B Cell:</u>	71.30-59.00
<u>Plasma-blasts:</u>	72.00-68.25
<u>Platelets:</u> as fragments of megakaryocytes with life span of 5-9 days, at death give off serotonin. *With depression, check need to increase norepinephrine or decrease noradrenaline.*	83.50-95.50
<u>Yellow Marrow:</u>	28.75-28.50
<u>Lymphoblasts:</u>	50.25-29.00
<u>Lymphocytes:</u>	43.25-34.00
<u>Metamyelocytes:</u>	33.20-26.50
<u>Monoblasts:</u>	68.25-80.50
<u>Monocytes:</u>	81.10-60.20
<u>Myeloblasts:</u>	67.25-42.50
<u>Myelocytes:</u>	65.25-48.25
<u>Polymorphonuclears:</u>	31.25-27.60
<u>Promyelocytes:</u>	52.75-40.75
Motile Phagocytic Cells:	13.25-15.00
<u>Blood Macrophages:</u>	21.20- 9.50

Urinary System

The urinary system's function is to filter blood and create urine as a waste by-product. Organs of the urinary system include the kidneys, renals, ureters, bladder, and urethra.

Bladder: 0.00-13.00, 8.00-67.00, 13.00-67.00, 26.25-20.50, 48.00-75.00

Also balance with 3.00-82.00.

- Arteries: 23.75-24.50
- Female: 19.00-14.00
- Helpful Paired with Kidney [3.00-82.00]: 39.00-30.00
  Pair with color black, emotion = fear, sound = groaning
- Ligaments: 22.00-19.40
- Lymph Vessels: 21.25-19.50
- Male: 21.00-15.00
- Mucous Coat: 21.50-24.25
- Muscular Coat: 41.75-25.00
  - External Longitudinal Layer: 35.10-24.25
  - Internal Longitudinal Layer: 26.75-23.00
  - Middle Circular Layer: 30.00-24.50
- Nerves, Parasympathetic: motor to bladder wall and inhibitory to the sphincter 25.20-26.75
- Nerves, Sympathetic: motor to sphincter and inhibitory to muscles of bladder 20.75-29.50
- Serous Coat: 29.00-26.00
- Sub-mucous Coat: 31.75-22.00
- Veins: 66.75-29.60

Calyces Minor: 31.60-42.75

Calyx of Kidney: 52.25-50.50

Kidneys: 23.00-58.00, 32.00-23.00, 0.00-23.00, 0.00-82.00

Check for balance of arsenic, DDT, earache, edema, flu, meningococcus, migraines, nausea, polio, poisons, scar tissue, and/or sniffles.

- Arteries: 53.50-33.50
- Cortical Substance: 42.25-37.50
- Fibrous Capsule: 38.00-72.50
- Helpful Paired with Bladder [39.00-30.00]: 3.00-82.00
  Pair with color black, emotion = fear, sound = groaning
- Lymph Vessels: 35.50-19.75
- Medullary Substance: 62.00-66.50
- Nerves: 33.75-22.75
- Renal Capsules: 43.25-57.50

Epithelial Lining:	41.50-10.75
Glomerular Capsule, Bowman's Capsule:	39.25-43.10
Glomerulus: one of the small structures in the Malpighian body of the kidney that is made up of capillaries in a cluster enveloped in a thin wall, giving off uriniferous tubules	46.00-65.25
Uriniferous Tubules:	53.75-19.75
Veins:	38.00-12.50
Pelvis of Ureter:	34.25-33.25
Urachus:	34.50-34.25
Ureters: the tubes that carry urine from the kidney to the bladder	0.00-67.00, 43.75-48.00
Left:	45.00-28.25
Right:	40.75-12.50
Arteries:	11.50-27.75
Fibrous Coat:	47.00-23.50
Mucous Coat:	27.25-25.50
Muscular Coat:	35.25-25.25
Nerves:	26.75-20.75
Urethra: a canal for discharge of urine from the bladder to outside the body	0.00-86.00, 25.50-13.25
Mucous Membrane:	28.25-31.60
Muscular Coat:	29.00-28.25
Skene's Glands: glands lying just inside and posterior to the female urethra	40.75-29.25
Sphincter Urethrae:	22.75-27.50
Spongy Erectile Tissue:	20.25-19.75

Other Systems

Breasts

The breasts are comprised of mammary glands, a compound alveolar gland with 15 to 20 lobes of glandular tissue separated from each other by interlobular septa. Each lobe is drained by a lactiferous duct which opens on the tip of the nipple. The mammary gland secretes milk for nourishment of the young. Males are born with a much smaller amount of breast tissue.

Mammary Glands:	0.00-26.00, 0.00-56.00, 12.50-56.00, 24.75-24.75, 59.00-92.00
Two compound glands of the breast that secrete milk.	
Arteries: carry oxygenated blood from the heart to the tissues	23.75- 7.50
From Intercostal Artery: between the ribs	24.10-19.50
From Internal Mammary Artery:	19.75-22.00
Thoracic Lateral: relating to the side of the chest or thorax	30.00-23.50
Fatty Tissue: *(analysis only - balance with 12.50-31.50)*	25.50-17.75
Fibrous Tissue:	17.75-98.25
Gland Tissue: tissue surrounding the gland	27.00-26.00
Areolar Tissue: circular tissue of different pigmentation	10.40- 7.25
Blood Vessels: the veins, arteries, and capillaries	9.75-22.40
Colostrum Corpuscles: breast fluid secreted two or three days after childbirth and before the onset of true lactation	8.75- 9.00
Ducts: tubular vessels or channels conveying secretions from a gland	33.60-25.50
Lactiferous Sinuses: secreting and conveying milk through a narrow opening	14.00-11.25
Lobes: well defined part of an organ separated by boundaries	17.75-11.75
Lobules: smaller lobes	22.75-23.00
Lymphatic Vessels: thin-walled vessels moving alkaline fluid from lymph tissues	29.60-39.50
Nerves: a bundle or group of bundled of nerve fibers from outside the central nervous system that connect the brain and spinal cord with the rest of the body	41.60-39.60
From 4th Thoracic: pertaining to the chest or thorax	50.75-52.50
From 5th Thoracic: pertaining to the chest or thorax	40.75-30.75
From 6th Thoracic: pertaining to the chest or thorax	33.90-26.90
Veins: vessels carrying dark red, unaerated blood to the heart	16.60-22.75
Nipple: the conical protuberance in each breast from which lactiferous ducts discharge	40.75-29.50
Areolar Glands: large, modified sweat glands lying beneath the areolar of the breast with ducts opening on the surface. They secrete lipoid material that lubricates the nipple	3.75- 9.50

Ductless Glands

These glands secrete one or more hormones that have a specific action on the body.

Parathyroid Gland: 0.00-92.00, 34.60-42.00, 86.00-4.00

Located alongside the thyroid gland. Secretes hormone important for controlling calcium-phosphorus balance. *If high tendency to clot, balance thyroid and parathyroid together.*

- Arteries: 31.50-26.00
- Connective Tissue: 52.75-36.50
  - Neutrophil Cells: 22.75-20.00
  - Oxyphil Granules: 9.00-30.00
  - Principal Cells: 40.25-38.50
- Lymph Cells: thin walled vessels conveying lymph from the tissues 64.50-53.50
- Nerves: 47.75-71.60
- Veins: 43.20-38.00

Pineal Body: a pine cone shaped, gland-like body in the brain 19.00-45.75

- Connective Tissue Cells: 35.25-35.50
- Fibrils: very small filamentous structures 34.75-30.00
- Pineal Cells: 76.00-47.75

Pituitary Body: 0.00-87.00, 6.00-87.00, 28.00- 9.75

With depression, check the need to increase enkephalins or serotonin.

- Anterior Lobe: 74.00-46.75, 74.00-65.00
  With depression, check the need to increase noradrenaline.
  - Alpha Cells: cells of the anterior lobe of the pituitary and the pancreas 52.00-32.50
  - Beta Cells: insulin-secreting cells of the pancreas that comprise the bulk of the islets of Langerhans 32.00-32.50
  - Chromophobe Cells: any of faintly staining cells of the anterior lobe of of the pituitary gland 46.25-43.40
- Pars Intermedia: 28.25-33.20
  - Blood Vessels: the veins, arteries, and capillaries 36.25-32.20
- Posterior Lobe: 56.00-23.00
  With depression check need to increase enkephalins, serotonin, affected by Lyme Disease.
  - Cell Columns: 45.00-35.75
  - Neuroglia Cells: non-nerve cells of the supporting tissue of the central nervous system and the retina of the eye 52.00-28.50
- Suprarenal Glands: 3.00- 9.00, 19.25-44.75
  Use for anemia, catarrh, dema, exhaustion, exhaustion, mononucleosis, multiple sclerosis, poisons, shock, tuberculosis, toxicity.

Cortex: 33.00-39.00, 47.50-40.25
Helps with cancer cell disturbance, also use for rheumatic, fever, heart, flu, and lagrippe.

- Zona Fasciculus: an acute inflammatory disease associated with small bundles of nerve or muscle fibers 40.75-35.00
- Zona Glomerulus: an acute inflammatory disease associated with one of the small structures of the Malpighian of the kidney, made up of capillary blood vessels in a cluster and enveloped in a thin wall 43.20-31.00
- Zona Reticular: an acute inflammatory disease associated with a form of connective tissue consisting of a network of reticular fibers and cells 32.00-25.00

Fibro Areolar Capsule: fibrous tissue in an areolar arrangement 76.50-71.50

Medulla: inner and central portions of an organ 28.00-23.00, 37.75-43.75

Arteries: 42.75-31.25

Chromaffin Cells: containing brown-staining granules 33.40-25.60

Venous Sinusoids: sinuses of the dura mater draining the brain and those of the spleen 44.00-30.25

Sympathetic Nerve Fibers: 35.00-36.25

Trabeculae: fibrous cord of connective tissue that serves as supporting fiber by forming a septum that extends into an organ from its wall or capsule 16.25-17.75

Thymus Gland: trains or modifies T-cells, which regulate the humoral immune system 0.00-77.00, 41.25-33.00, 59-77.00

Areolar Tissue: loose connective tissue that occupies the interspace of the body 47.25-35.50

Arteries: 36.75-29.50

Concentric Corpuscles: 29.50-31.50

Lobules: small lobes 44.75-38.50

Veins: 42.50-32.00

Thyroid Gland: a gland of internal secretion 0.00-11.00

Base of the neck on both sides: 0.00-14.50

Of the lower part of the larynx: 0.00-98.00

Upper part of the trachea: 6.00-81.00, 57.00-98.00, 82.00-51.00

Capsule Connective: 36.00-29.00

Isthmus: narrow structure connecting the two larger halves 39.00-35.00

Left Lobe: 45.50-44.00

Pyramidal Lobe: 61.50-25.75

Right Lobe: 36.25-38.50

Vesicles: 32.00-39.00

Ears

Ears, Inner

Arteries of the Labyrinth: osseous and membranous labyrinths that are essential to maintaining physical equilibrium and balance	61.00-57.75
Internal Auditory Artery:	59.00-54.50
Stylomastoids Branch of Auricular Artery:	76.50-76.75
Auditory Nerve: 8th cranial nerve, a sensory nerve with two sets of fibers	62.20-52.50
Cochlear Nerve: division of the nerve that supplies the cochlea	33.50-63.75
Spiral Ganglion: long-coiled ganglion in the cochlea of the ear	57.50-57.00
Vestibular Ganglion: a bi-lobed ganglion located in the vestibular branch	53.80-55.60
Vestibular Nerve: includes the superior, inferior, and middle branches	64.60-59.90
Bony Labyrinth: includes vestibule, three semicircular canals, and cochlea	68.50-46.00
Bony Labyrinth Canals:	49.25-43.25
Cochlea: winding, cone-shaped tube in the inner ear	41.50-54.50
• Helicotrema: opening at the top of the cochlea canal where the scala tympani and scala vestibuli unite	40.25-52.80
• Modiolus: central pillar of axial part of cochlea extending from the base to the apex	54.50-44.75
• Osseous Spiral Lamina: a thin bony plate projecting from the modiolus into the cochlea canal dividing into the scala tympani and scala vestibuli	48.00-44.00
• Scala Tympani: the cochlear duct filled with perilymph found below the spiral lamina	35.50-43.00
• Scala Vestibuli: the cochlear duct forming the upper portion of the osseous canal	31.00-41.50
• Secondary Spiral Lamina:	40.75-46.60
• Spiral Canal: the bony cochlea enclosing the scala tympani, scala vestibuli, and cochlear duct	46.25-46.00
Fibro-Serous Membrane Lining:	38.50-49.00
Perilymph: pale, limpid fluid contained between the membranous and bony labyrinth of the internal ear	43.75-43.00
Pyramid of the Vestibule:	61.50-55.60
Semicircular Canals: superior, inferior, and posterior passages of the inner ear	48.75-50.50
• Ampullae:	46.80-35.40
• Crus Commune:	47.00-47.00
• Posterior Semicircular Canal:	57.00-52.75
• Superior Semicircular Canal:	57.00-48.40

Vestibule: small cavity at the beginning of the canal	45.50-40.40
• Vestibular Crest:	60.00-50.25
Membranous Labyrinth:	41.60-36.60
Duct of Cochlea:	70.50-60.25
• Basilar Membrane:	60.50-58.25
• Lagena:	49.75-53.40
• Limbus Laminae Spiralis:	38.50-28.50
• Stria Vasularis:	50.75-55.50
• Vestibular Membrane:	55.60-42.75
Ductus Ultriculosaccular: pertains to the utricle and saccule of the labyrinth	75.00-54.50
Endolymph: pale transparent fluid within the labyrinth of the ear	48.00-45.50
Hair Cells:	54.25-43.75
Macula of the Utricle:	44.20-54.75
• Filaments Auditory Nerve:	42.25-57.50
Rods of Corti:	48.50-54.50
• Inner:	46.50-81.75
• Outer:	58.00-64.50
Saccule: smaller of two sacs compromising the portion of the membranous labyrinth occupying the vestibule of the inner ear	83.40-78.75
• Canalis Reuniens:	48.00-70.00
• Filament Auditory Nerve:	50.00-36.00
• Macula of the Saccule:	18.20-57.60
• Saccus Endolymphatic:	49.20-57.25
Semicircular Ducts:	63.00-52.60
• Ampulla: sac-like dilation of a canal or duct	67.75-45.75
• Crista Ampullaris: localized thickening of the membrane lining the ampullae of the semicircular canals; covered with neuroepithelium containing auditory cells	70.10-60.60
• Septum Transversum:	76.25-73.25
Spiral Organ of Corti: structure in floor of cochlear duct resting on basilar membrane that contains the hair cell receptors of stimuli produced by sound	46.50-71.00
Tectorial Membrane: top of the membrane	46.50-49.50
Utricle: one of the two sacs of the membranous labyrinth in the bony vestibule of the inner ear	40.75-51.20
Wax or Cerumen: the wax-like brown secretions in the external canal of the ear. *Balance 40.00-6.00 for excessive secretions.*	4.00-6.00, 68.75-75.00

Ear, Middle

Auditory Ossicles: the bones of the middle ear	49.00-62.75
Articulations of Ossicles: the union between the small bones of the ear, such as the malleus, incus, and stapes	48.80-49.00
Incus: the middle of the three ossicles in the tympanum and the anvil	41.50-49.60
• Ligaments of Malleus:	62.10-47.20
Stapes: ossicle in the middle ear that articulates with the incus (stirrup)	57.00-44.40
• Ligaments of Stapes:	46.75-49.00
Pharyngotympanic, Auditory, and Eustachian Tubes:	45.75-48.75
Bony Part:	48.20-60.25
Cartilaginous Part:	59.00-55.20
Mucous Membrane: membrane lining passages and cavities communicating with the air. Consists of a surface layer of epithelium, a basement membrane, and the lamina - an underlying layer of connective tissue.	51.40-50.20
Tube Tonsil:	51.50-55.25

Eyes

Eyeball 0.00-18.00, 32.00-18.00, 53.25-53.00

The globe of the eye has three humors aqueous, the lens crystalline, and vitreous. *Also balance with 9.50-10.50.*

Aqueous Humor: transparent liquid contained in the anterior and posterior chambers of the eye, which is produced by the ciliary processes from the anterior chamber and then to the venous system by way of the canal of Schlemm 41.00-56.00

Ciliary Body: thickened part of the vascular tunic of the eye between the base of the iris and the anterior part of the choroid. Has three zones: disk, crown and muscle. 51.40-56.00

Ciliary Muscle: alters shape of crystalline lens to accommodate near vision 39.00-41.50

Ciliary Processes: secretes fluids that nourish the cornea, lens, and vitreous body. Also serves as attachment points for the lens' suspensory ligaments. 51.00-48.00

Ciliary Ring: 46.50-47.00

Eyelids: 51.20-64.60

- Conjunctiva: mucous membranes that line eyelids and is reflected onto the eyeballs 0.00-19.00, 56.00-22.00
- Lymph Glands: 33.75-35.60
- Superior Tarsus: a curved plate of dense, white, fibrous tissue that forms the supporting structure of the eyelid 34.50-39.00

Fibers of Lens: 72.50-57.25

Fibrous Coat: 55.00-47.50

- Cornea: the clear, transparent anterior portion of the fibrous coat of the eye, comprising about one sixth of the surface 24.00-33.50, 60.10-54.25
- Corneal Epithelium: 55.75-46.75
- Posterior Elastic Lamina: 85.50-54.40
  - Pectinate Ligament of Iris: 61.50-49.25
- Substantia Propria: 46.50-44.50
  - Corneal Spaces: 44.00-51.00
  - Lamellae: 48.25-49.50
- Sclera: 22.50-37.00, 50.75-41.50
  - Ciliary Nerves: 51.25-59.00
  - External Surface: 55.50-47.50
  - Retina, Optic Chiasma: 0.00-22.50
- Tunica Albuginea: the white fibrous coat of the eye 26.75-20.25

Iris: 0.00-27.50, 0.00-28.00, 48.20-56.50

- Arteries: 80.25-69.00
- Layer of Endothelial Cells: 75.00-62.50

Muscular Fibers:	55.75-58.50
Nerves:	50.75-61.20
• Ciliary Ganglion Branches:	63.75-65.10
• From Oculomotor Nerve:	57.75-68.20
• Long Ciliary:	51.00-49.50
Pupil:	27.00-25.00, 78.60-80.75
Stroma:	66.20-55.50
Lacrimal Apparatus:	61.20-64.20
Ducts:	0.00-64.00, 60.50-59.25
Gland:	43.50-56.60
Sac:	56.40-61.60
Lens: 50.00-38.25	
Capsule:	57.25-56.50
Nucleus:	56.50-68.50
Orbital Cavity:	35.75-45.75
Orbital Muscles:	50.60-46.25
Rectus:	
• Inferior:	60.00-27.50
• Lateral:	47.25-48.00
• Medial:	14.00-20.00
• Oblique Inferior:	33.00-43.20
• Oblique Superior:	56.50-56.50
• Superior:	26.75-56.00
Retina:	40.50-54.30, 48.00-54.00
Retina, Optic Chiasma:	0.00-22.50
Arteries:	48.50-50.25
Ganglionic Layer:	46.00-50.50
Inner Nuclear Layer:	42.60-53.80
Inner Plexiform Layer:	52.40-53.00
Layer of Rods and Cones:	33.20-49.50
Macula Lutea:	59.00-55.00
Outer Nuclear Layer:	44.00-59.00
Outer Plexiform Layer:	40.00-43.10
Stratum Opticum:	46.00-39.00

Vascular Coat: composed of blood vessels 31.50-41.50

Choroid: 0.00-36.50, 25.00-37.00, 70.00-55.50
dark brown vascular coat of the eye between the sclera and retina, extending from ora serrata to optic nerve; consists of blood vessels united by connective tissue containing five layers of pigmented cells.

- Basal Lamina: 48.00-46.00
- Capillaries: 50.25-48.75
- Choriocapillary or Choriocapillaris Lamina: middle layer of the choroid containing close mesh of capillaries 46.50-45.40
- Suprachoroid or Suprachoroidea Lamina: outermost layer of the choroid 64.75-65.30
- Vascular Lamina: thin flat layer of membrane composed of blood vessels 43.00-57.00

Vitreous Body: a transparent jellylike mass that fills the cavity of the eyeball, enclosed by the hyaloid membrane 32.75-43.20

Hyaloid Canal: lymph channel in vitreous extending from optic desk to posterior capsule of lens 56.00-55.25

Hyaloid or Hyaloiditis Membrane: envelops the vitreous humor 34.00-32.20

Suspensory Ligament of Lens: supporting structure 61.00-55.20

Hair, Nails, and Skin

- **Hair:** 80.25-58.20
  - Cortex of Hair: 58.75-65.75
  - Hair Follicle: 0.00-85.00, 36.60-46.00
    - Henle's Layer: 47.00-40.25
    - Hyaline Layer of Dermic Coat: 64.50-63.20
    - Outer Layer of Dermic Coat: 33.60-46.00
  - Medulla of Hair: 78.50-66.00
- **Nails:** 60.50-72.50
  - Glomus: circulation 62.50-64.75
  - Lymphatics: 52.75-48.25
  - Matrix: 46.20-55.50
  - Nail Bed: 65.50-55.10
  - Nail Plate: 49.50-55.25
  - Nail Root: 42.75-59.00
  - Nail Wall: 54.75-59.50
- **Skin:**
  - Alveoli: 38.50-45.00
    - Basement Membrane: 39.25-49.50
    - Epithelial Cells: 40.00-44.50
    - Larger Cells Containing Fat: 42.00-36.50
    - Sebum Cutaneous or Debris: 34.40-41.00
  - Corium: 33.00-46.00
    - Arteries: 57.00-49.50
    - Connective Tissue: 40.50-41.00
      - Papillary Layer: 48.50-49.50
      - Reticular Layer: 43.00-41.00
    - Ducts: 36.50-42.50
    - Lymph Vessels: 50.75-52.20
    - Muscular Fibers: 58.40-46.50
    - Nerves: 34.00-28.25
    - Sebaceous Glands: 44.10-42.00
    - Veins: 45.50-53.00
  - Epidermis: 0.00-42.00, 70.00-2.50, 91.00-31.00
    - Germinative Zone: 47.25-46.00
      - Basal Cell Layer: 56.00-49.50
      - Prickle Cell Layer: 42.00-37.40
    - Horny Layer: 59.60-52.50
      - Clear Layer: 35.40-61.00
      - Granular Layer: 50.50-33.00
      - Horny Layer: 79.25-56.50
  - Epithelial Cells: 49.40-53.00
  - Pigment, Melanin: 58.20-53.20
  - Sweat Glands: 26.40-39.75

Energetic Diseases and Disorders

Acid, Uric: crystalline deposits of this mineral may cause arthritis	4.00-8.00, 12.50-20.50
Adhesion, Pleural:	24.75-28.50
Allergies, Hay Fever: inflammation to air passages induced by external irritation	8.00-80.00
Arteries:	
Aneurysm: often with radiating pain, sac formed by local enlargement of a weakened wall of an artery	31.00-35.50
Arteriosclerosis: thickening of artery walls and loss of elasticity	21.00-47.00
Arteritis: inflammatory disorder of the arteries	28.50-23.50
Arthritis: inflammation of a joint accompanied by pain	10.00-9.00, 13.00-29.00
Degenerative: also known as osteoarthritis	9.75-19.00
Rheumatoid: chronic, systematic disease with inflammatory changes in joints changes in joints and related structures, and crippling deformities	17.50-19.00
Stiffness:	4.00- 7.00
Bacterial:	
Chorea/Saint Vitus' Dance: nervous system disorder with jerking movements	30.75-19.50
Mycobacterial Tuberculosis: *Also balance Adrenals [3.00-9.00].*	42.00-41.50
Blood Clot: coagulated blood mass. *Also balance thyroid and parathyroid.*	53.50-41.50
Blood Pressure, High: an increase in tonus or tension	16.75-26.00
Blood Pressure, Low: a deficiency in tonus or tension	28.25-16.00
Bruises: an injury appearing as an area of discolored skin on the body, caused by a blow or impact rupturing underlying blood vessels.	17.75-18.50
Calculus, Stones: any abnormal concretion within the body, usually comprised of mineral salts. Most often found in the bladder, kidneys, ureter, or urethra.	46.75-34.00
Carcinoma, Cells: abnormal, malignant growth or tumors in epithelial tissue	30.00-scan
Carcinoma, Toxins: malignant tumors	30.00- 0.00
Cartilage, Displaced: indicated by pain and increased movement in the joint	39.50-33.50
Chill: an attack of shivering, coldness, and/or pallor of the skin produced by involuntary contraction of many muscle groups	53.00-31.50
Compression:	39.25-30.50
Concussion: brain injury caused by a blow to the head	22.25-16.50
Congenital Predisposition:	25.25-33.00
Congestion: excess blood or fluid in a tissue or organ	4.00-8.00, 4.00-81.00, 40.00-91.00
Contraction: shortening/tightening of a muscle	19.25-62.00
Coronary Thrombosis: occlusion of one or more coronary arteries	44.25-44.50
Cramps: painful, spasmodic contraction of one or more muscles	23.60-10.00
Degeneration: deterioration of cells, organs, tissues, or related structures	11.25-41.50

Dermatitis: skin inflammation evidenced by itching, redness and/or lesions	26.75-50.50
Diet, Improper:	67.00-27.00
Edema: local or generalized fluid retention in tissues. Also called dropsy. *Also balance Adrenals [3.00-9.00] and Pituitary [56.00-23.00] as needed, then 3.00-82.00, 6.00-17.00, and 23.00-23.00.*	28.75-31.50
Effusion: escape of fluid into a tissue or organ	25.00-29.50
Emphysema: abnormal inflation of the alveoli (tiny air sacs) in the lungs, damaging their walls and making it harder to breathe	32.75-12.00
Empyema: pus-filled pockets in the body cavity	16.00-14.50
Endocarditis: inflammation of the lining membrane of the heart	42.25-44.00
Enlargement:	33.00-34.00
Exostosis: bony growth arising from a bone	22.00-32.00
Fatty Degeneration: a change involving deposition of fat in the cytoplasm	43.25-33.75
Fibroid, Fibrositis: a fibrous, encapsulated connective tissue tumor	33.75-31.00
Heart, Angina Pectoris: condition marked by recurrent sharp, radiating, referred pain in the chest and left arm caused by a decrease in blood supply to the heart muscle. *Also balance with 20.00-41.00.*	62.00-30.00
Heart, Cardiac Asthma: asthma secondary to heart disease	26.25-9.00
Septicemia: bacteria in the blood. *Also balance with colloidal silver.*	8.75-35.00

External Trauma:

Burns: tissue damage from chemical, electrical, radioactive, or thermal energy	20.50-17.75
Burns, Thermal: tissue damage from contact with fire, hot fluids, or hot objects	73.00-41.50
Electrical Shock:	17.50-15.00
Insulin Shock: from overdosage of insulin	7.75-12.50
Physical Shock: sudden drop in blood pressure	9.25-13.00
Poisoning Shock:	12.50-34.25
Radioactive Fallout:	13.25-13.50
Shock Therapy:	10.75-14.50
Surgical Shock:	15.50-19.75
Traumatic Shock: due to accidents, injury, or surgery	21.75-17.20

Gastrointestinal Disorders:

Appendicitis: inflammation of the appendix, often with sharp pain	44.50-22.00
Colitis, Large Intestine: inflammation of the colon	40.00-91.00
Ulcers, Stomach: *For pain increased by food, balance 5.00-91.00 & Prostate*	92.00-85.00

Hepatic and Biliary Disorders:

Addiction: including alcoholism and drug addiction	28.75-26.00
Calculus, Biliary: abnormal stony mass in bile tract	7.00-35.00, 39.25-38.00
Hernia: a protrusion of an organ through the wall or cavity that normally contains it. *Balance 59.00-17.00, 31.00-2.00, 5.00-91.00, then 6.00-81.00, then 77.00-41.00. Also check Staph.*	27.25-23.50, 40.00-41.00
Hyperinsulinism: An excess amount of insulin in the blood	28.50-31.75
Infection, Fungus, Bacteria, Actinomycosis: results in bony degeneration of jaws (Lumpy Jaw), abscesses in lungs and intestines	32.25-37.50
Innervation, Impaired: impaired function of the nervous system	32.50-17.00
Lesion: a circumscribed area of pathologically altered tissue	24.50- 5.00
Lymphadenoma: hyperplasia (pre-cancer, or cancer) of the lymph nodes	25.25-25.50
Metabolic Imbalance: imbalance between physical and chemical processes	8.25-29.25
Muscular Atrophy/Dystrophies: wasting away or failure to develop body tissue	16.50-15.50
Muscular Strain:	22.00-10.75
Osteoarthritis: degenerative wearing away of cartilage and connective tissue	23.25-23.25
Nerves, Impinged:	21.80-23.50
Neuralgia: sharp pain along the course of a nerve	71.50-71.00
Otorhinolaryngological Disorders: nasal and paranasal disorders, catarrh *Also balance Adrenals [3.00-9.00].*	39.50-33.50
Catarrh, Chronic: inflammation of nasal and throat membranes	18.25-18.30
Pain, Muscle Rupture: increases with movement or pressure	28.75-23.50
Pain, Periostitis: increases at night	10.75-14.50
Pain, Rheumatism: increases with movement	37.75-44.00
Paralysis Agitans: rigidity marked by weakness, delay of voluntary motion, a peculiar festinating gait, and muscular contraction	11.00-19.00
Perforation: process of making a hole, as with ulceration	56.00-53.00
Pericardial Effusion: escape of fluid into the pericardium	23.00- 6.75
Pericarditis: inflammation of the pericardium	47.25-46.25
Peritoneal Effusion: escape of fluid into the peritoneum	23.00-19.25
Phlebitis: inflammation of the vein(s)	23.00-28.50
Pleural Effusion: escape of fluid into the pleura	23.25-32.75
Psychological Shock:	19.50-20.00
Emotional:	19.25-19.50
Financial:	19.75-20.00
Mental Oppression:	19.75-18.75
Sexual:	18.50-11.25
Pulmonary Disorders:	
Asthma:	40.00-18.50
Bronchitis:	22.00-17.00

Pleural Disorders: pleurisy, often with sharp pain, relieved by pressure increased by movement, breathing, or coughing	25.25-41.75
Pneumonia: inflammation of the lungs caused by infection or irritants	32.25-11.00
Radiation, Actinobacillosis: bacilli infection caused by radiation	31.75-32.50
Rheumatism, Chronic: joint disorders resulting in deformity of the joint, including rheumatoid arthritis, gout, and degenerative diseases	17.25-35.00
Rheumatism, Gonorrheal: arthritis resulting from gonorrheal infection	23.25-18.00
Skeletal Dislocation:	3.00-80.00, 18.20-15.60
Skeletal Fracture:	12.00-12.00
Spasm: painful, involuntary contraction of the muscle(s)	33.25- 3.50
Sprain: trauma to a joint marked by pain and inflammation	28.00-48.50
Stenosis: constriction or narrowing of a passage	28.00-47.25
Subluxation: partial or incomplete separation	10.50- 4.50
Synovial Effusion: escape of fluid into the synovial membrane	19.25-36.00
Synovitis: inflammation of the synovial membrane surrounding the joints	8.60-39.00
Throat: including the pharynx, adenoids, and adenoid area	22.00-26.25
Thrombosis: formation or existence of a blood clot within the vascular system	36.50-25.50
Toxic State:	32.20-21.00
Trauma: *Also balance 68.00-49.00 after surgery, accident, or other trauma*	10.25-24.00
Tumor: a swelling of a part of the body, generally without inflammation, caused by an abnormal growth of tissue, whether benign or malignant _	
Benign: non-recurrent swelling or enlargement	28.00-27.50
Cyst: a closed sac or pouch that contains fluid, semifluid or solid material	25.00-28.00
Fibroma: a tumor of fibrous, encapsulated, connective tissue	27.25-11.00
Fibroma, Uterine: a fibroid tumor of the uterus	38.75-27.00
Malignant: cancerous tumors with cells that grow uncontrollably and spread locally and/or to distant sites.	57.20-40.50
Polypus: a tumor with a pedicle	49.25-14.00
Sarcoma: a form of cancer that develops from connective tissue	26.25-26.50
Ulcers: open sore or lesion of the skin or mucous membrane accompanied by sloughing of inflamed necrotic tissue	44.50-17.00
Ulcers, Varicose:	21.75-37.50
Valvular Incompetence: inability of a bodily valve to close completely	36.50-60.25
Viral Diseases, Respiratory:	22.00- 2.00
Warts: circumscribed cutaneous elevation of the epidermis caused by the papillomavirus. *Also balance acid/alkaline balance and sodium chloride.*	10.75-20.50

Patterns of Information	Rate Left	Rate Right	Energetic Processes
Acid, Uric, Crystals (Delawarr)	04.00	08.00	Deposit of this Mineral Could Cause Arthritic Problems, Balance 5.00-91.00, Check Gout 19.50-33.75 or 85.00-56.00
Acid, Uric, Crystals (Delawarr)	12.50	20.50	
Adhesions, Pleural (Delawarr)	24.75	28.50	Freq. 43.0
Arthritis (Delawarr)	17.50	16.00	Balance 4.00-7.00 For Stiffness, Balance also 10.00-9.00, 13.00-29.00, Adrenals, colon, Kidneys & Thyroid; Freq: 43.3
Arthritis, Degenerative (Delawarr)	09.75	19.00	Check For Mineral Deposits, Calcium, Carbon-Stop Massive Carbohydrate Intake, Silicon, Sodium-Use Sun-Dried Sea Salt, Uric Acid; Freq: 43.3
Arthritis, Rheumatoid (Delawarr)	17.50	19.00	Check For Mineral Deposits, Calcium, Carbon-Stop Massive Carbohydrate Intake, Silicon, Sodium-Use Sun-Dried Sea Salt, Uric Acid; Freq: 43.3
Blood, Blood Corpuscles (Delawarr)	14.00	36.20	
Blood, Blood Corpuscles, Hemoglobin, (Delawarr)	49.75	22.75	
Blood, Blood Corpuscles, Lymph Corpuscles, (Delawarr)	52.30	54.25	
Blood, Blood Corpuscles, Plasma Cell, (Delawarr)	55.60	61.75	With Depression Check Need to Decrease Enkephalins, Melatonin, Serotonin
Blood, Blood Corpuscles, Plasma Salts, (Delawarr)	63.75	67.50	With Depression Check Need To Increase Norepinephrins, Or Decrease Enkephalins, Melatonin, Serotonin
Blood, Blood Corpuscles, Platelets, (Delawarr)	58.75	48.10	With Depression Check Need To Increase Norepinephrins, Or Decrease Noradrenaline
Blood, Blood Corpuscles, Red Corpuscles, Erythrocyte, (Delawarr)	51.20	41.30	
Blood, Blood Corpuscles, Reticulocyte, (Delawarr)	35.00	29.75	
Blood, Blood Corpuscles, White Corpuscles, Eosonophils, (Delawarr)	70.00	75.00	If higher than GV an indicator of parasite or allergies
Blood, Blood Corpuscles, White Corpuscles, Leucocytes, (Delawarr)	59.00	42.00	
Blood, Blood Corpuscles, White Corpuscles, Lymphocytes, (Delawarr)	52.25	49.00	
Blood, Blood Corpuscles, White Corpuscles, Lymphocytes, Basophils, (Delawarr)	32.60	56.75	
Blood, Blood Corpuscles, White Corpuscles, Macrophages, (Delawarr)	65.00	59.00	
Blood, Blood Corpuscles, White Corpuscles, Monocytes, (Delawarr)	23.75	25.50	
Blood, Blood Corpuscles, White Corpuscles, Polymorphonuclears, (Delawarr)	37.25	56.75	
Blood, Blood Corpuscles, White Corpuscles, Polymorphonuclears, Neutrophils, (Delawarr)	69.50	69.00	
Blood, Blood Corpuscles, White Corpuscles, Polymorphonuclears, Oxyphils, (Delawarr)	51.60	28.00	
Blood, Clot, Hematoma, (Delawarr)	53.50	41.50	Check Thyroid/Parathyroid Balance
Blood, Hemostatic Agents, Fibrinogen, (Delawarr)	65.00	64.50	A Protein that helps form blood clots, High Elevations Indicate Hidden Risk for Heart Disease
Blood, Hemostatic Agents, Heparin, (Delawarr)	65.75	59.00	
Blood, Hemostatic Agents, Thrombokinase, (Delawarr)	64.60	40.00	
Blood, Hemostatic Agents, Thromboplastin, (Delawarr)	30.75	55.25	
Blood, Hemostatic Agents, Tissue Fibrinogen, (Delawarr)	61.50	69.00	A Protein that helps form blood clots, High Elevations Indicate Hidden Risk for Heart Disease
Blood, Homeostatic Agents, Fibrin, (Delawarr)	40.00	32.40	A Whitish, filamentous Protein formed by the action of thrombin on fibrinogen, entangles red & white blood cells & platelets forming a clot
Blood, Homeostatic Agents, Prothrombin, (Delawarr)	49.00	50.50	
Blood, Homeostatic Agents, Thrombin, (Delawarr)	45.00	55.40	
Blood, Plasma, (Delawarr)	31.20	19.00	
Blood, Plasma, General Vitality, (Delawarr)	09.00	49.00	Use With Caution, With Depression Check Need To Decrease Enkephalins, Melatonin, Serotonin
Blood, Plasma, Volume, Organ, (Delawarr)	00.00	49.00	Note Eb#, balance 9.00-49.00 with caution. Freq. 321.9
Blood, Pressure, High, (Delawarr)	16.75	26.00	Check Structure, Emotions, Hormones, Boron

Patterns of Information	Rate Left	Rate Right	Energetic Processes
Blood, Pressure, Low, (Delawarr)	28.25	16.00	
Breast, Mammary Glands, (Delawarr)	00.00	56.00	
Breast, Mammary Glands, (Delawarr)	00.00	26.00	Amino Acid, Carnitine should be high in human breast milk
Breast, Mammary Glands, (Delawarr)	24.75	24.75	
Breast, Mammary Glands, (Delawarr)	59.00	92.00	
Breast, Mammary Glands, (Delawarr)	12.50	56.00	
Breast, Mammary Glands, Arteries, (Delawarr)	23.75	07.50	
Breast, Mammary Glands, Arteries, From Intercostal Artery, (Delawarr)	24.10	19.50	
Breast, Mammary Glands, Arteries, From Internal Mammary Artery, (Delawarr)	19.75	22.00	
Breast, Mammary Glands, Arteries, Thoracic Lateral, (Delawarr)	30.00	23.50	Same as Abundance
Breast, Mammary Glands, Fatty Tissue, (Delawarr)	25.50	17.75	Use 12.50-31.50
Breast, Mammary Glands, Fibrous Tissue, (Delawarr)	17.75	98.25	
Breast, Mammary Glands, Gland Tissue, (Delawarr)	27.00	26.00	
Breast, Mammary Glands, Gland Tissue, Areolar, (Delawarr)	10.40	07.25	
Breast, Mammary Glands, Gland Tissue, Blood Vessels, (Delawarr)	09.75	22.40	
Breast, Mammary Glands, Gland Tissue, Colostrum Corpuscles, (Delawarr)	08.75	09.00	
Breast, Mammary Glands, Gland Tissue, Ducts, (Delawarr)	33.60	25.50	
Breast, Mammary Glands, Gland Tissue, Lactiferous Sinuses, (Delawarr)	14.00	11.25	
Breast, Mammary Glands, Gland Tissue, Lobes, (Delawarr)	17.75	11.75	
Breast, Mammary Glands, Gland Tissue, Lobules, (Delawarr)	22.75	73.00	
Breast, Mammary Glands, Lymph Vessels, (Delawarr)	29.60	39.50	
Breast, Mammary Glands, Nerves, (Delawarr)	41.60	39.60	
Breast, Mammary Glands, Nerves, From 4th Thoracic, (Delawarr)	50.75	52.50	
Breast, Mammary Glands, Nerves, From 5th Thoracic, (Delawarr)	40.75	30.75	
Breast, Mammary Glands, Nerves, From 6th Thoracic, (Delawarr)	33.90	26.90	
Breast, Mammary Glands, Nipple, (Delawarr)	40.75	29.50	
Breast, Mammary Glands, Nipple, Areolar Glands, (Delawarr)	03.75	09.50	
Breast, Mammary Glands, Veins, (Delawarr)	16.60	22.75	
Breast, Mammary Glands, Veins, Circulus Venosus, (Delawarr)	15.60	68.25	
Bruises, (Delawarr)	17.75	18.50	
Calculus, Stones, (Delawarr)	46.75	34.00	
Carcinoma, Cells, Abnormal Growth, (Delawarr)	30.00	100.00	Freq. 2127.0, 43.3
Carcinoma, Toxins, Malignant Tumor, (Delawarr)	30.00	00.00	Freq. 2127.0, 43.3
Carcinoma, Toxins, Malignant Tumor, (Delawarr)	30.00	00.00	Freq. 666.0, 690.0, 740.0, 1840.0, 1998.0, 2000.0
Cartilage, Displaced, Pain, Increased by Movement, (Delawarr)	39.50	33.50	
Cell Structure, Protoplasm, (Delawarr)	50.00	46.60	
Cell Structure, Protoplasm, Cytocentrum, Attraction Sphere, (Delawarr)	64.25	73.00	
Cell Structure, Protoplasm, Cytocentrum, Centrioles, (Delawarr)	59.00	57.10	
Cell Structure, Protoplasm, Cytoplasm, (Delawarr)	53.20	26.00	
Cell Structure, Protoplasm, Cytoplasm, Chromophil Granules, (Delawarr)	45.00	60.50	
Cell Structure, Protoplasm, Cytoplasm, Golgi's Network, (Delawarr)	75.50	75.50	

Patterns of Information	Rate Left	Rate Right	Energetic Processes
Cell Structure, Protoplasm, Cytoplasm, Hyaloplasm, (Delawarr)	55.75	64.50	
Cell Structure, Protoplasm, Cytoplasm, Mitochrondria, In Brown Fat Cells, (Delawarr)	54.50	39.00	
Cell Structure, Protoplasm, Cytoplasm, Tonofibrils, (Delawarr)	69.75	55.50	
Cell Structure, Protoplasm, Nucleus, (Delawarr)	47.60	66.00	
Cell Structure, Protoplasm, Nucleus, Chromatin Granules, (Delawarr)	64.10	65.30	
Cell Structure, Protoplasm, Nucleus, Chromosomes, (Delawarr)	55.50	52.00	
Cell Structure, Protoplasm, Nucleus, Chromosomes, X-Chromosomes, (Delawarr)	60.75	56.00	Ovaries, Ovum, Genes Of Combined with
Cell Structure, Protoplasm, Nucleus, chromosomes, Y-Chromosomes, (Delawarr)	50.00	44.50	
Cell Structure, Protoplasm, Nucleus, Genes, (Delawarr)	54.50	56.20	
Cell Structure, Protoplasm, Nucleus, Nuclear Membrane, (Delawarr)	47.50	49.00	
Cell Structure, Protoplasm, Nucleus, Nucleolus, (Delawarr)	50.50	31.50	
Central Nervous System, (Delawarr)	05.00	62.00	
Central Nervous System, Autonomic Nervous System, (Delawarr)	40.40	42.50	
Central Nervous System, Autonomic Nervous System, Parasympathetic System, (Delawarr)	57.00	62.50	Nerve Fiber
Central Nervous System, Autonomic Nervous System, Parasympathetic System, (Delawarr)	50.00	42.00	Hormone, Histamine
Central Nervous System, Autonomic Nervous System, Sympathetic System, (Delawarr)	05.00	61.50	
Central Nervous System, Autonomic Nervous System, Sympathetic System, (Delawarr)	05.00	60.00	
Central Nervous System, Autonomic Nervous System, Sympathetic System, Ganglia, (Delawarr)	35.00	37.50	
Central Nervous System, Autonomic Nervous System, Sympathetic System, Plexuses, (Delawarr)	45.25	60.10	
Central Nervous System, Brain Areas, Association Area, (Delawarr)	53.50	80.75	
Central Nervous System, Brain Areas, Association Area, Frontal, (Delawarr)	00.00	82.75	
Central Nervous System, Brain Areas, Association Area, Frontal, (Delawarr)	00.00	05.00	Note Eb#, With Depression Check Need to Increase Enkephalins, Melatonin, Noradrenaline, Serotonin Or Decrease Dopamine. Freq. 315.8
Central Nervous System, Brain Areas, Association Area, Frontal, (Delawarr)	37.00	26.00	
Central Nervous System, Brain Areas, Association Area, Post Antral, (Delawarr)	41.60	29.60	With Depression Check Need To Increase Noradrenaline, Serotonin, Or Decrease Dopamine, Area Of Memory
Central Nervous System, Brain Areas, Auditory Area, Temporal Lobe, (Delawarr)	39.25	65.50	
Central Nervous System, Brain Areas, Balance & Equilibrium, Cerebellar Cortex, (Delawarr)	48.25	43.00	
Central Nervous System, Brain Areas, Coordination Of Movement, Cerebellar Cortex, (Delawarr)	57.25	40.75	
Central Nervous System, Brain Areas, Postrolandic Area, (Delawarr)	23.50	25.00	
Central Nervous System, Brain Areas, Postrolandic Area, Arm, (Delawarr)	40.40	35.40	
Central Nervous System, Brain Areas, Postrolandic Area, Body-Face, (Delawarr)	35.75	14.50	
Central Nervous System, Brain Areas, Postrolandic Area, Leg, (Delawarr)	32.50	25.00	
Central Nervous System, Brain Areas, Postrolandic Area, Trunk, (Delawarr)	80.50	63.75	

Patterns of Information	Rate Left	Rate Right	Energetic Processes
Central Nervous System, Brain Areas, Prerolandic Area, (Delawarr)	63.00	48.25	
Central Nervous System, Brain Areas, Prerolandic Area, Arm, (Delawarr)	58.50	52.50	
Central Nervous System, Brain Areas, Prerolandic Area, Body-Face, (Delawarr)	52.50	28.50	
Central Nervous System, Brain Areas, Prerolandic Area, Eyes & Head, (Delawarr)	38.00	37.50	Freq. 20.0
Central Nervous System, Brain Areas, Prerolandic Area, Leg, Body-Thighs, (Delawarr)	40.00	53.50	
Central Nervous System, Brain Areas, Prerolandic Area, Trunk, (Delawarr)	48.75	46.25	
Central Nervous System, Brain Areas, Sensory Area, Smell, (Delawarr)	82.20	60.50	Check Hypocampus, Bi-Polar
Central Nervous System, Brain Areas, Sensory Area, Taste, (Delawarr)	60.00	55.00	
Central Nervous System, Brain Areas, Sensory Area, Uncinate Gyrus, (Delawarr)	39.50	39.00	
Central Nervous System, Brain Areas, Sex Areas, Desire, (Delawarr)	96.25	15.00	
Central Nervous System, Brain Areas, Sex Areas, Endocrine Center, (Delawarr)	34.75	52.00	
Central Nervous System, Brain Areas, Sex Areas, Function, (Delawarr)	47.50	51.00	
Central Nervous System, Brain Areas, Sex Areas. Frustration, (Delawarr)	69.50	71.00	Base, Anterior Auditory Artery
Central Nervous System, Brain Areas, Sex, Frigidity, (Delawarr)	43.00	54.00	
Central Nervous System, Brain Areas, Sleep Center, (Delawarr)	43.50	43.50	
Central Nervous System, Brain Areas, Speech, Broca's Area, (Delawarr)	42.75	47.25	
Central Nervous System, Brain Areas, Visual Area, Occipital, (Delawarr)	41.25	46.00	
Central Nervous System, Brain Areas, Visuo-Psychic Area, (Delawarr)	45.50	45.00	
Central Nervous System, Brain, Fore, Pineal Gland, (Delawarr)	15.75	28.25	With Depression Check Need to Increase Melatonin, Noradrenaline, Serotonin Or Decrease Dopamine, Will Be Low for Nervous Conditions; Freq: 20.0, 5000.0
Central Nervous System, Cranial Nerves - 01st Pair, Olfactory, (Delawarr)	28.40	28.50	from the epithelial lining of the nose, entirely sensory, sense of smell, many travel to the temporal lobe of the cerebral cortex
Central Nervous System, Cranial Nerves - 02nd Pair, Optic, (Delawarr)	44.50	18.50	Carry Impulses Associated With Vision, Entirely Sensory, Originates In Retina Of Eyes, Nerve Fibers Can Terminate In The Thalamus Or The Visual Cortex In Occipital Lobe
Central Nervous System, Cranial Nerves - 03rd Pair, Oculomotor, (Delawarr)	43.60	43.50	From midbrain & travels to the 6 extrinsic muscle or superior rectus, medial rectus, inferior oblique, These nerves supply levator palpebrae superioris muscles that raise the upper child
Central Nervous System, Cranial Nerves - 04th Pair, Trochlear, (Delawarr)	48.75	34.20	Small Nerves that carry impulses to & from the superior oblique muscles, helps with the voluntary movements of the eyeball
Central Nervous System, Cranial Nerves - 05th Pair, Trigeminal, (Delawarr)	43.00	42.00	Come from the Pons, Major sensory nerves of face, contain the impulses for muscles for chewing including masseter & temporalis
Central Nervous System, Cranial Nerves - 06th Pair, Abducens, (Delawarr)	37.50	38.00	Involved With The Coordinated Contraction Of Extrinsic Muscles Of Both Eyes, Synchronization Of Nerve Impulses From 3Rd, 4Th & 5Th
Central Nervous System, Cranial Nerves - 07th Pair, 7th Facial Nerve, (Delawarr)	49.50	33.75	Contain sensory neurons from taste buds of the tongue & supply motor nerves of face & scalp, carry neurons to the tear glands & salivary glands
Central Nervous System, Cranial Nerves - 08th Pair, Vestibulocochlear, (Delawarr)	41.75	29.50	Transmit impulses of hearing & equilibrium
Central Nervous System, Cranial Nerves - 09th Pair, Glossopharyngeal, (Delawarr)	39.25	38.50	Supplies muscle of pharynx (swallowing), carry impulses that are sensitive to changes in level of blood oxygen & carbon dioxide in carotid sinus, neurons carry impulses from taste buds in back of tongue, membranes of pharynx, tonsils & mid-ear cavity

Patterns of Information	Rate Left	Rate Right	Energetic Processes
Central Nervous System, Cranial Nerves - 10th Pair, Vagus Nerve, (Delawarr)	39.50	27.50	Arising in Medulla, Oblongata & Providing Parasympathetic Innervation to Larynx, Lungs, Heart, Esophagus & abdominal organs, regulates heart rate, depth of respiration, digestive process
Central Nervous System, Cranial Nerves - 11th Pair, Spinal Accessory, (Delawarr)	27.00	33.40	Nerves that supply the traezius & sternocleidomastoid muscles
Central Nervous System, Cranial Nerves - 12th Pair, Hypoglossal, (Delawarr)	39.75	45.00	Located beneath the tongue & supply the muscles of the tongue
Central Nervous System, Cranial Nerves, (Delawarr)	33.50	35.75	
Central Nervous System, Cranial Nerves, Optic, (Delawarr)	22.00	82.00	
Central Nervous System, Cranial Nerves, Spinal Or Peripheral Nerves, (Delawarr)	31.75	27.75	
Central Nervous System, Cranial Nerves, Spinal Or Peripheral Nerves, Cervical 8 Pairs, (Delawarr)	01.00	21.00	
Central Nervous System, Cranial Nerves, Spinal Or Peripheral Nerves, Cervical 8 Pairs, (Delawarr)	31.75	30.75	
Central Nervous System, Cranial Nerves, Spinal Or Peripheral Nerves, Coccygeal 1 Pair, (Delawarr)	41.25	32.50	
Central Nervous System, Cranial Nerves, Spinal Or Peripheral Nerves, Lumbar 5 Pairs, (Delawarr)	35.50	18.50	
Central Nervous System, Cranial Nerves, Spinal Or Peripheral Nerves, Sacral 5 Pairs, (Delawarr)	40.50	28.25	
Central Nervous System, Cranial Nerves, Spinal Or Peripheral Nerves, Thoracic 12, (Delawarr)	33.50	30.25	Pairs
Central Nervous System, Ductless Glands, Pineal Body, (Delawarr)	00.00	27.00	With depression check need to increase melatonin, noradrenaline, serotonin or decrease dopamine, will be low for nervous conditions. Freq. 20.0, 5000.0
Central Nervous System, Fore Brain, (Delawarr)	00.00	78.50	With Depression Check Need to Increase Serotonin, Main Area of Brain
Central Nervous System, Fore Brain, (Delawarr)	57.75	87.50	With Depression Check Need to Increase Serotonin, Main Area of Brain
Central Nervous System, Fore Brain, Cerebral Cortex White Matter, (Delawarr)	79.00	81.00	With Depression Check Need To Increase Enkephalins, Norepinephrins, Serotonin Or Decrease Dopamine, Noradrenaline
Central Nervous System, Fore Brain, Corpus Callosum, (Delawarr)	50.00	38.00	With Depression Check Need To Increase Enkephalins, Norepinephrines, Serotonin Or Decrease Dopamine, Melatonin
Central Nervous System, Fore Brain, Hypothalamus, (Delawarr)	00.00	90.50	
Central Nervous System, Fore Brain, Hypothalamus, Cowpers Glands, (Delawarr)	00.00	02.00	Should Equal GV, Check 22.00-0.00, 77.00-0.00, Balance Higher With Care
Central Nervous System, Fore Brain, Hypothalamus, Nerves, (Delawarr)	53.00	50.50	
Central Nervous System, Fore Brain, Hypothalamus, Nerves, (Delawarr)	00.00	51.50	Balance 22.00-0.00, 77.00-0.00, Use Higher With Care
Central Nervous System, Fore Brain, Lateral Ventricles, (Delawarr)	61.50	62.00	
Central Nervous System, Fore Brain, Thalamus, (Delawarr)	00.00	54.50	With Depression Check Need To Increase Norepinephrines In
Central Nervous System, Fore Brain, Thalamus, (Delawarr)	52.50	40.00	With Depression Check Need To Increase Norepinephrines In
Central Nervous System, Fore Brain, Thalamus, Caudata Nucleus, (Delawarr)	45.00	28.00	With Depression Check Need To Increase Melatonin, Norepinephrines, Serotonin
Central Nervous System, Fore Brain, Thalamus, Internal Capsule, (Delawarr)	34.75	33.20	With Depression Check Need To Increase Norepinephrines In
Central Nervous System, Fore Brain, Thalamus, Lentiform Nucleus, (Delawarr)	47.00	31.75	With Depression Check Need To Decrease Enkephalins, Serotonin
Central Nervous System, Fore Brain, Thalamus, To Improve Memory, Nerves, (Delawarr)	40.50	34.50	With Depression Check Need To Increase Enkephalins, Melatonin, Norepinephrines, Serotonin Or Decrease Dopamine, Use With Hippocampus 32.5-42 & With Acetylcholine Hormone 39.5-36.25
Central Nervous System, Fore Brain, Third Ventricle, (Delawarr)	65.25	58.00	
Central Nervous System, Hind Brain, (Delawarr)	80.75	72.00	Main Area of Brain
Central Nervous System, Hind Brain, Cerebellum, (Delawarr)	00.00	43.25	With Depression Check Need To Decrease Enkephalins Area of Emotion
Central Nervous System, Hind Brain, Cerebellum, Dentate Nucleus, (Delawarr)	57.25	39.75	With Depression Check Need To Decrease Enkephalins Area of Emotion

Patterns of Information	Rate Left	Rate Right	Energetic Processes
Central Nervous System, Hind Brain, Cerebellum, Fourth Ventricle, (Delawarr)	19.50	10.00	With Depression Check Need To Decrease Enkephalins Area of Emotion
Central Nervous System, Hind Brain, Cerebellum, Fourth Ventricle, (Delawarr)	00.00	04.50	With Depression Check Need To Decrease Enkephalins
Central Nervous System, Hind Brain, Cerebellum, Grey Matter, (Delawarr)	44.50	38.00	With Depression Check Need To Decrease Enkephalins Area of Emotion
Central Nervous System, Hind Brain, Cerebellum, Inferior Peduncle, (Delawarr)	22.75	21.00	With Depression Check Need To Decrease Enkephalins Area of Emotion
Central Nervous System, Hind Brain, Cerebellum, Left, (Delawarr)	02.00	29.00	With Depression Check Need To Decrease Enkephalins Area Of Emotions
Central Nervous System, Hind Brain, Cerebellum, Middle Peduncle, (Delawarr)	41.50	47.25	With Depression Check Need To Decrease Enkephalins Area of Emotion
Central Nervous System, Hind Brain, Cerebellum, Nerves, Epileptogenic Zone, (Delawarr)	57.75	46.75	With Depression Check Need To Decrease Enkephalins Area of Emotion
Central Nervous System, Hind Brain, Cerebellum, Superior Peduncle, (Delawarr)	21.50	21.25	With Depression Check Need To Decrease Enkephalins Area of Emotion
Central Nervous System, Hind Brain, Cerebellum, White Matter, (Delawarr)	54.00	58.75	With Depression Check Need To Decrease Enkephalins Area of Emotion
Central Nervous System, Hind Brain, Medulla Oblongata, (Delawarr)	28.00	13.00	
Central Nervous System, Hind Brain, Medulla Oblongata, Olivary Nucleus, (Delawarr)	57.00	83.50	
Central Nervous System, Hind Brain, Nerves, Cerebellum, Right, (Delawarr)	02.00	31.00	With Depression Check Need To Decrease Enkephalins Area of Emotion
Central Nervous System, Hind Brain, Pons Varolii, Fore, Medulla, (Delawarr)	00.00	15.00	With Depression Check Need To Decrease enkephalins
Central Nervous System, Hind Brain, Pons, Mid, Medulla, (Delawarr)	00.00	10.00	Should Equal GV, With Depression Check Need To Decrease enkephalins
Central Nervous System, Mid Brain, (Delawarr)	30.25	30.50	With Depression Check Need to Decrease Dopamine, Main Area of Brain, Area of Emotions
Central Nervous System, Mid Brain, Cerebral Peduncles, (Delawarr)	39.25	19.00	
Central Nervous System, Mid Brain, Cerebral Peduncles, Aqueduct Of Mid Brain, (Delawarr)	41.75	16.50	
Central Nervous System, Mid Brain, Cerebral Peduncles, Basis Pedunculi, (Delawarr)	39.50	32.00	
Central Nervous System, Mid Brain, Cerebral Peduncles, Corpora Quadrigemina, (Delawarr)	32.10	32.50	
Central Nervous System, Mid Brain, Cerebral Peduncles, Red Nucleus, (Delawarr)	43.25	38.25	
Central Nervous System, Mid Brain, Cerebral Peduncles, Substantia Nigra, (Delawarr)	44.50	26.50	With Depression Check Need To Increase Melatonin, Or Decrease Serotonin
Central Nervous System, Mid Brain, Cerebral Peduncles, Tegmentum, (Delawarr)	37.75	35.00	
Central Nervous System, Spinal Cord, Anterior Horns, (Delawarr)	44.50	51.00	
Central Nervous System, Spinal Cord, Coccyx Balance, (Delawarr)	40.50	22.00	
Central Nervous System, Spinal Cord, Extradural Space, (Delawarr)	48.25	47.75	
Central Nervous System, Spinal Cord, Grey Matter, (Delawarr)	44.75	39.50	
Central Nervous System, Spinal Cord, Grey Matter, Nerve Cells, (Delawarr)	73.20	46.20	
Central Nervous System, Spinal Cord, Grey Matter, Nerve Fibers, (Delawarr)	48.75	68.50	
Central Nervous System, Spinal Cord, Grey Matter, Neuroglia, (Delawarr)	71.50	71.00	
Central Nervous System, Spinal Cord, Meninges, Arachnoid Mater, (Delawarr)	48.75	40.40	
Central Nervous System, Spinal Cord, Meninges, Arachnoid Villi, (Delawarr)	03.25	46.25	
Central Nervous System, Spinal Cord, Meninges, Cerebrospinal Fluid, (Delawarr)	68.25	57.50	With Depression Check Need To Increase Enkephalins, Noradrenaline, Norepinephrins, Serotonin Or Decrease Dopamine, Melatonin

Patterns of Information	Rate Left	Rate Right	Energetic Processes
Central Nervous System, Spinal Cord, Meninges, Cerebrospinal Fluid, Choroid Plexuses, (Delawarr)	64.25	45.50	With Depression Check Need To Increase Enkephalins, Melatonin, Norepinephrins, Serotonin Or Decrease Dopamine
Central Nervous System, Spinal Cord, Meninges, Dura Mater, (Delawarr)	59.50	52.50	
Central Nervous System, Spinal Cord, Meninges, of Brain & Spinal Cord, (Delawarr)	27.00	09.00	
Central Nervous System, Spinal Cord, Meninges, of Brain & Spinal Cord, (Delawarr)	58.00	82.00	
Central Nervous System, Spinal Cord, Meninges, of Brain & Spinal Cord, Nerves, (Delawarr)	00.00	50.50	
Central Nervous System, Spinal Cord, Meninges, Pia Mater, (Delawarr)	87.00	34.00	
Central Nervous System, Spinal Cord, Subarachnoid Space, (Delawarr)	44.75	40.20	
Central Nervous System, Spinal Cord, Subdural Space, (Delawarr)	35.50	73.00	
Central Nervous System, Spinal Cord, White Matter, (Delawarr)	70.50	28.75	
Central Nervous System, Spinal Cord, White Matter, Medullated Nerve Fibers, (Delawarr)	59.00	39.00	
Central Nervous System, Spinal Cord, White Matter, Medullated Nerve Fibers, Anterior, (Delawarr)	46.75	44.00	Column
Central Nervous System, Spinal Cord, White Matter, Medullated Nerve Fibers, Lateral, (Delawarr)	48.50	38.00	Column
Central Nervous System, Spinal Cord, White Matter, Medullated Nerve Fibers, Posterior, (Delawarr)	49.60	37.50	Column
Chill, (Delawarr)	53.00	31.50	
Circulatory Disorders, Arteries, Arteriosclerosis, (Delawarr)	21.00	47.00	A Thickening of the Walls of the Arteries, Loss of Elasticity. Freq. 43.34
Circulatory Disorders, Arteries, Arteritis, Inflammatory Disorder of the Arteries, (Delawarr)	28.50	23.50	
Circulatory Disorders, Artery, Aneurysm, Often With Radiating Pain, (Delawarr)	31.00	35.50	Sac Formed by Local Enlargement of a Weakened Wall of the Artery
Circulatory System, Arteries, (Delawarr)	81.25	78.00	
Circulatory System, Arteries, (Delawarr)	68.00	71.00	
Circulatory System, Arteries, (Delawarr)	00.00	71.00	
Circulatory System, Arteries, Internal Coat, (Delawarr)	67.00	59.50	
Circulatory System, Arteries, Middle Coat, (Delawarr)	28.75	29.50	
Circulatory System, Arteries, Nerves, (Delawarr)	46.60	54.00	
Circulatory System, Arteries, Nerves, From Peripheral Nerves, (Delawarr)	61.75	55.25	
Circulatory System, Arteries, Nerves, From Perivascular Plexuses, Splanchnic Arteries, (Delawarr)	70.25	55.25	
Circulatory System, Arteries, Nerves, From Plexuses, (Delawarr)	58.40	76.20	
Circulatory System, Arteries, Nerves, From Sympathetic Ganglia, (Delawarr)	39.50	84.00	
Circulatory System, Arteries, Outer Coat, (Delawarr)	32.50	30.75	
Circulatory System, Arterioles, (Delawarr)	73.25	40.75	Parts of the Arteries
Circulatory System, Arterioles, Arteriole Capillaries, (Delawarr)	32.75	41.60	
Circulatory System, Capillaries, (Delawarr)	66.75	44.25	
Circulatory System, Capillaries, Cells Of Capillary Wall, (Delawarr)	39.00	50.00	
Circulatory System, Capillaries, Cells Of Envelope, (Delawarr)	59.25	52.50	

Patterns of Information	Rate Left	Rate Right	Energetic Processes
Circulatory System, Veins, (Delawarr)	00.00	97.00	
Circulatory System, Veins, (Delawarr)	50.00	40.75	
Circulatory System, Veins, (Delawarr)	00.00	69.00	
Circulatory System, Veins, External Coat, (Delawarr)	61.00	73.10	
Circulatory System, Veins, Internal Coat, (Delawarr)	40.75	31.75	
Circulatory System, Veins, Middle Coat, (Delawarr)	50.50	66.50	
Circulatory System, Veins, Nerves, (Delawarr)	87.25	43.00	
Circulatory System, Veins, Valves, (Delawarr)	71.50	57.20	
Compression, (Delawarr)	39.25	30.50	
Concussion, (Delawarr)	22.25	16.50	
Congenital, Predisposition, (Delawarr)	25.25	33.00	
Congestion, (Delawarr)	40.00	08.60	Use 4.00-81.00, 4.00-8.00, 40.00-91.00
Contraction, (Delawarr)	19.25	62.00	
Coronary, Thrombosis, Blood Clot, (Delawarr)	44.25	44.50	
Cramps, (Delawarr)	23.60	10.00	Freq. 26.0
Degeneration, (Delawarr)	11.25	41.50	
Dermatitis, (Delawarr)	26.75	50.50	
Detox (3) Delawarr, Lymphatic System, Support & Stimulate All Subtle Fields of	44.00	41.00	Detox Support Program, Affected By Lyme Disease; Freq. 10.0, 880.0
Diet, Improper, (Delawarr)	67.00	27.00	
Diseases, Bacterial, Cholera, Saint Vitus' Dance, Nervous System Disorder, (Delawarr)	30.75	19.50	Characterized by Jerking or Irregular Movements, Check out for Lingering Symptoms
Diseases, Mycobacterial, Tuberculosis, (Delawarr)	42.00	41.50	Use Of 3.00-9.00 Adrenals is Effective. Freq. 800.0, 43.3
Diseases, Septicemia, Bacteria in the Blood, (Delawarr)	08.75	35.00	Colloidal Silver is Effective
Disorders, Heart, Angina Pectoris, Conditon Marked by Recurrent Sharp, Radiating,, (Delawarr)	62.00	30.00	Referred Pain in the Chest & Left Arm; Use Also 20.00-41.00, Decrease in Blood Supply to Heart Muscle
Disorders, Heart, Cardiac Asthma, (Delawarr)	26.25	09.00	Freq. 43.3
Ductless Glands, Parathyroid Gland, (Delawarr)	86.00	04.00	If High Tendency to Clot, Balance Thyroid with Parathyroid, Located Near Or Alongside the Thyroid gland, Secretes Hormone Important In Controlling Calcium-Phosphorus Balance
Ductless Glands, Parathyroid Gland, (Delawarr)	34.60	42.00	If High Tendency to Clot, Balance Thyroid with Parathyroid, Located Near Or Alongside the Thyroid gland, Secretes Hormone Important In Controlling Calcium-Phosphorus Balance
Ductless Glands, Parathyroid Gland, (Delawarr)	00.00	92.00	
Ductless Glands, Parathyroid Gland, Arteries, (Delawarr)	31.50	26.00	
Ductless Glands, Parathyroid Gland, Connective Tissue, (Delawarr)	52.75	36.50	
Ductless Glands, Parathyroid Gland, Connective Tissue, Neutrophil Cells, (Delawarr)	22.75	20.00	
Ductless Glands, Parathyroid Gland, Connective Tissue, Oxyphil Granules, (Delawarr)	09.00	30.00	
Ductless Glands, Parathyroid Gland, Connective Tissue, Principal Cells, (Delawarr)	40.25	38.50	
Ductless Glands, Parathyroid Gland, Lymph Vessels, (Delawarr)	64.50	53.50	
Ductless Glands, Parathyroid Gland, Nerves, (Delawarr)	47.75	71.60	
Ductless Glands, Parathyroid Gland, Veins, (Delawarr)	43.20	38.00	
Ductless Glands, Pineal Body, (Delawarr)	19.00	45.75	Freq. 20.0, 5000.0
Ductless Glands, Pineal Body, (Delawarr)	16.00	28.00	With Depression Check Need to Increase Melatonin, Noradrenaline, Serotonin Or Decrease Dopamine, Will Be Low for Nervous Conditions. Freq. 20.0, 5000.0
Ductless Glands, Pineal Body, Connective Tissue Cells, (Delawarr)	35.25	35.50	Freq. 20.0, 5000.0
Ductless Glands, Pineal Body, Fibrils, (Delawarr)	34.75	30.00	Freq. 20.0, 5000.0
Ductless Glands, Pineal Body, Pineal Cells, (Delawarr)	76.00	47.75	Freq. 20.0, 5000.0

Patterns of Information	Rate Left	Rate Right	Energetic Processes
Ductless Glands, Pituitary Body, (Delawarr)	00.00	87.00	With Depression Check Need to Increase Enkephalins, Serotonin
Ductless Glands, Pituitary Body, (Delawarr)	06.00	87.00	
Ductless Glands, Pituitary Body, (Delawarr)	28.00	09.75	
Ductless Glands, Pituitary Body, Anterior Lobe, (Delawarr)	74.00	46.75	With Depression Check Need To Increase Noradrenaline
Ductless Glands, Pituitary Body, Anterior Lobe, (Delawarr)	74.00	65.00	With Depression Check Need To Increase Noradrenaline
Ductless Glands, Pituitary Body, Anterior Lobe, Alpha Cells, (Delawarr)	52.00	32.50	
Ductless Glands, Pituitary Body, Anterior Lobe, Beta Cells, Produce Insulin, (Delawarr)	32.00	32.50	
Ductless Glands, Pituitary Body, Anterior Lobe, Chromophobe Cells, (Delawarr)	46.25	43.40	
Ductless Glands, Pituitary Body, Pars Intermedia, (Delawarr)	28.25	33.20	
Ductless Glands, Pituitary Body, Pars Intermedia, Blood Vessels, (Delawarr)	36.25	32.20	
Ductless Glands, Pituitary Body, Posterior Lobe, (Delawarr)	66.25	55.75	With Depression Check Need to Increase Enkephalins, Serotonin, Affected by Lyme Disease
Ductless Glands, Pituitary Body, Posterior Lobe, (Delawarr)	56.00	23.00	With Depression Check Need to Increase Enkephalins, Serotonin, Affected by Lyme Disease
Ductless Glands, Pituitary Body, Posterior Lobe, Cell Columns, (Delawarr)	45.00	35.75	
Ductless Glands, Pituitary Body, Posterior Lobe, Neuroglia Cells, (Delawarr)	52.00	28.50	
Ductless Glands, Suprarenal (Adrenal) Glands, (Delawarr)	03.00	09.00	Balance For Anemia, Catarrh, Edema, Exhaustion, Mononucleosis, Multiple Sclerosis, Poisons, Shock, Tuberculosis, Toxicity
Ductless Glands, Suprarenal (Adrenal) Glands, (Delawarr)	19.25	44.75	
Ductless Glands, Suprarenal (Adrenal) Glands, Cortex, (Delawarr)	47.50	40.25	
Ductless Glands, Suprarenal (Adrenal) Glands, Cortex, (Delawarr)	33.00	39.00	Helpful With Reducing Cancer Cell Disturbance, Use For Rheumatic Fever, Heart, Flu & Lagrippe
Ductless Glands, Suprarenal (Adrenal) Glands, Cortex, Zona Fasciculata, (Delawarr)	40.75	35.00	
Ductless Glands, Suprarenal (Adrenal) Glands, Cortex, Zona Glomerulosa, (Delawarr)	43.20	31.00	
Ductless Glands, Suprarenal (Adrenal) Glands, Cortex, Zona Reticularis, (Delawarr)	32.00	25.00	
Ductless Glands, Suprarenal (Adrenal) Glands, Fibro-areolar Capsule, (Delawarr)	76.50	71.50	
Ductless Glands, Suprarenal (Adrenal) Glands, Medulla, (Delawarr)	28.00	23.00	
Ductless Glands, Suprarenal (Adrenal) Glands, Medulla, (Delawarr)	37.75	43.75	
Ductless Glands, Suprarenal (Adrenal) Glands, Medulla, Arteries, (Delawarr)	42.75	31.25	
Ductless Glands, Suprarenal (Adrenal) Glands, Medulla, Chromaffin Cells, (Delawarr)	33.40	25.60	
Ductless Glands, Suprarenal (Adrenal) Glands, Medulla, Venous Sinusoids, (Delawarr)	44.00	30.25	
Ductless Glands, Suprarenal (Adrenal) Glands, Sympathetic Nerve Fibers, (Delawarr)	35.00	36.25	
Ductless Glands, Suprarenal (Adrenal) Glands, Trabeculae, (Delawarr)	46.25	47.70	
Ductless Glands, Thymus Gland, , Areolar Tissue, (Delawarr)	47.25	35.50	
Ductless Glands, Thymus Gland, Arteries, (Delawarr)	36.75	29.50	
Ductless Glands, Thymus Gland, Concentric Corpuscles, (Delawarr)	29.50	31.50	
Ductless Glands, Thymus Gland, Lobules, (Delawarr)	44.75	38.50	

Patterns of Information	Rate Left	Rate Right	Energetic Processes
Ductless Glands, Thymus Gland, Veins, (Delawarr)	42.50	32.00	
Ductless Glands, Thyroid Gland, (Delawarr)	00.00	14.50	Freq. 20.0, 160.0
Ductless Glands, Thyroid Gland, (Delawarr)	82.00	51.00	Freq. 20.0, 160.0
Ductless Glands, Thyroid Gland, (Delawarr)	06.00	81.00	Use to Eliminate 82.00-XX. Freq. 20.0, 160.0
Ductless Glands, Thyroid Gland, (Delawarr)	57.00	98.00	Freq. 20.0, 160.0
Ductless Glands, Thyroid Gland, (Delawarr)	00.00	98.00	Freq. 20.0, 160.0
Ductless Glands, Thyroid Gland, Capsule of Connective Tissue, (Delawarr)	36.00	29.00	Freq. 20.0, 160.0
Ductless Glands, Thyroid Gland, Isthmus, (Delawarr)	39.00	35.00	Freq. 20.0, 160.0
Ductless Glands, Thyroid Gland, Left Lobe, (Delawarr)	45.50	44.00	Freq. 20.0, 160.0
Ductless Glands, Thyroid Gland, Organ, (Delawarr)	00.00	11.00	Freq. 20.0, 160.0
Ductless Glands, Thyroid Gland, Pyramidal Lobe, (Delawarr)	61.50	25.75	Freq. 20.0, 160.0
Ductless Glands, Thyroid Gland, Right Lobe, (Delawarr)	36.25	38.50	Freq. 20.0, 160.0
Ductless Glands, Thyroid Gland, Vesicles, (Delawarr)	32.00	39.00	Freq. 20.0, 160.0
Duodenum, (Delawarr)	16.00	24.50	Use 16.00-25.00 if Insect Stings, Migraines, Parathion, Poisons (Arsenic), D D T, Acidity in Burns, or Lindane is found in
Duodenum, (Delawarr)	00.00	25.00	Balance 16.00-25.00 if insect stings, migraines, parathion, poisons (arsenic), DDT, acidity in burns, or lindane is found in the duodenum
Duodenum, Arteries, (Delawarr)	34.25	43.00	
Duodenum, Lymph Vessels, (Delawarr)	22.75	21.25	
Duodenum, Mucous Membrane, (Delawarr)	26.25	42.25	
Duodenum, Mucous Membrane, Circular Folds, (Delawarr)	17.20	28.00	
Duodenum, Mucous Membrane, Duodenal Glands, (Delawarr)	28.25	41.00	
Duodenum, Mucous Membrane, Villi, (Delawarr)	22.25	17.00	
Duodenum, Mucous Membrane, Villi, Artery Leaving Villus, (Delawarr)	24.50	26.25	
Duodenum, Mucous Membrane, Villi, Capillaries, (Delawarr)	27.00	25.00	
Duodenum, Mucous Membrane, Villi, Columnar Epithelium, (Delawarr)	11.10	12.50	
Duodenum, Mucous Membrane, Villi, Goblet Cells, (Delawarr)	22.25	25.10	
Duodenum, Mucous Membrane, Villi, Lacteals, (Delawarr)	30.25	30.50	
Duodenum, Mucous Membrane, Villi, Lymph Vessels, (Delawarr)	36.60	27.00	
Duodenum, Mucous Membrane, Villi, Muscle Cells, (Delawarr)	25.00	27.00	Freq. 20.0, 240.0
Duodenum, Muscular Coat, (Delawarr)	43.25	22.50	
Duodenum, Muscular Coat, Circular Fibers, (Delawarr)	39.50	50.25	
Duodenum, Muscular Coat, Longitudenal Fibers, (Delawarr)	63.50	60.20	
Duodenum, Nerves, from Splanchnic Nerve, (Delawarr)	35.00	21.75	
Duodenum, Nerves, from Vagus Nerve, (Delawarr)	41.25	36.60	
Duodenum, Nerves, Meissner's Plexus, Vagal Origin, (Delawarr)	29.75	33.75	
Duodenum, Nerves, Myenteric Plexus, Vagal Origin, (Delawarr)	25.60	23.60	
Duodenum, Nerves, Sympathetics, Inhibit Peristalsis, (Delawarr)	22.25	21.75	
Duodenum, Peritoneal Covering, (Delawarr)	23.00	21.75	
Duodenum, Sub Mucous Coat, (Delawarr)	34.75	48.75	
Duodenum, Sub Mucous Coat, Blood Vessels, (Delawarr)	36.50	55.50	

Patterns of Information	Rate Left	Rate Right	Energetic Processes
Duodenum, Sub Mucous Coat, Lymphatics, (Delawarr)	38.10	55.25	
Duodenum, Sub Mucous Coat, Nerves, (Delawarr)	35.50	37.40	
Duodenum, Suspensory Muscle, (Delawarr)	34.25	52.25	Freq. 20.0, 240.0
Duodenum, Veins, (Delawarr)	31.50	31.25	
Edema, Fluid or Water Retention, (Delawarr)	28.75	31.50	Use Of 3.00-9.00 Adrenals is Effective, Check Posterior Pituitary 56.00-23.00, Use 6.00-17.00, 3.00-82.00, 23.00-23.00
Effusion, Escape of Fluid into a Part, (Delawarr)	25.00	29.50	
Emphysema, (Delawarr)	32.75	12.00	
Empyema, Pus in a Body Cavity, (Delawarr)	16.00	14.50	
Endocarditis, (Delawarr)	42.25	44.00	
Enlargement, (Delawarr)	33.00	34.00	
Exostosis, Bony Growth Arising from the Bone, (Delawarr)	22.00	32.00	
External & Physical Agent Disorders, Burns, (Delawarr)	20.50	17.75	
External & Physical Agent Disorders, Burns, Thermal, Shock, (Delawarr)	73.00	41.50	
External & Physical Agent Disorders, Electric Shock, (Delawarr)	17.50	15.00	
External & Physical Agent Disorders, Insulin Shock, (Delawarr)	07.75	12.50	
External & Physical Agent Disorders, Physical Shock, (Delawarr)	09.25	13.00	
External & Physical Agent Disorders, Poisoning Shock, (Delawarr)	12.50	34.25	
External & Physical Agent Disorders, Radioactive Fallout, Atomic, (Delawarr)	13.25	13.50	Potassium Chloride, Kelp Helpful
External & Physical Agent Disorders, Shock Therapy, Thermal, (Delawarr)	10.75	14.50	
External & Physical Agent Disorders, Surgical Shock, (Delawarr)	15.50	19.75	
External & Physical Agent Disorders, Traumatic Shock, (Delawarr)	21.75	17.20	
Eyes, Eyeball, (Delawarr)	53.25	53.00	Freq. 20.0
Eyes, Eyeball, (Delawarr)	00.00	18.00	Balance 9.50-10.50, often used for dreaming, Freq. 20.0
Eyes, Eyeball, (Delawarr)	32.00	18.00	Freq. 20.0
Eyes, Eyeball, Aqueous Humour, (Delawarr)	41.00	56.00	Freq. 20.0
Eyes, Eyeball, Ciliary Body, (Delawarr)	51.40	56.00	Freq. 20.0
Eyes, Eyeball, Ciliary Body, Ciliaris Muscle, Alters Shape of Crystalline Lens In, (Delawarr)	39.00	41.50	Accommodation. Freq. 20.0, 240.0
Eyes, Eyeball, Ciliary Body, Ciliary Processes, (Delawarr)	51.00	48.00	Freq. 20.0
Eyes, Eyeball, Ciliary Body, Ciliary Ring, (Delawarr)	46.50	47.00	Freq. 20.0
Eyes, Eyeball, Eyelids, (Delawarr)	51.20	64.60	Freq. 20.0
Eyes, Eyeball, Eyelids, Conjunctiva, (Delawarr)	56.00	22.00	Freq. 20.0
Eyes, Eyeball, Eyelids, Conjunctiva, (Delawarr)	00.00	19.00	Freq. 20.0
Eyes, Eyeball, Eyelids, Lymph Glands, (Delawarr)	33.75	35.60	Freq. 20.0
Eyes, Eyeball, Eyelids, Superior Tarsus, (Delawarr)	34.50	39.00	Freq: 20.0
Eyes, Eyeball, Fibers Of Lens, (Delawarr)	72.50	57.25	Freq: 20.0
Eyes, Eyeball, Fibrous Coat, (Delawarr)	55.00	47.50	Freq: 20.0
Eyes, Eyeball, Fibrous Coat, Cornea, (Delawarr)	60.10	54.25	Freq: 20.0
Eyes, Eyeball, Fibrous Coat, Cornea, (Delawarr)	24.00	33.50	Freq: 20.0
Eyes, Eyeball, Fibrous Coat, Cornea, Corneal Epithelium, (Delawarr)	55.75	46.75	Freq: 20.0
Eyes, Eyeball, Fibrous Coat, Cornea, Posterior Elastic Lamina, (Delawarr)	85.50	54.40	Freq: 20.0
Eyes, Eyeball, Fibrous Coat, Cornea, Posterior Elastic Lamina, Pectinate Ligament Of, (Delawarr)	61.50	49.25	Iris; Freq: 20.0
Eyes, Eyeball, Fibrous Coat, Cornea, Substantia Propria, (Delawarr)	46.50	44.50	Freq: 20.0
Eyes, Eyeball, Fibrous Coat, Cornea, Substantia Propria, Corneal Spaces, (Delawarr)	44.00	51.00	Freq: 20.0

Patterns of Information	Rate Left	Rate Right	Energetic Processes
Eyes, Eyeball, Fibrous Coat, Cornea, Substantia Propria, Lamellae, (Delawarr)	48.25	49.50	Freq: 20.0
Eyes, Eyeball, Fibrous Coat, Sclera, (Delawarr)	22.50	37.00	Freq: 20.0
Eyes, Eyeball, Fibrous Coat, Sclera, (Delawarr)	50.75	41.50	Freq: 20.0
Eyes, Eyeball, Fibrous Coat, Sclera, Ciliary Nerves, (Delawarr)	51.25	59.00	Freq: 20.0
Eyes, Eyeball, Fibrous Coat, Sclera, External Surface, (Delawarr)	55.50	47.50	Freq: 20.0
Eyes, Eyeball, Fibrous Coat, Sclera, Retina, Optic Chiasma, (Delawarr)	00.00	22.50	Freq. 20.0
Eyes, Eyeball, Iris, (Delawarr)	00.00	28.00	Freq. 20.0
Eyes, Eyeball, Iris, (Delawarr)	48.20	56.50	Freq: 20.0
Eyes, Eyeball, Iris, (Delawarr)	00.00	27.50	Freq. 20.0
Eyes, Eyeball, Iris, Arteries, (Delawarr)	80.25	69.00	Freq: 20.0
Eyes, Eyeball, Iris, Layer Of Endothelial Cells, (Delawarr)	75.00	62.50	Freq: 20.0
Eyes, Eyeball, Iris, Muscular Fibers, (Delawarr)	55.75	58.50	Freq: 20.0
Eyes, Eyeball, Iris, Nerves, (Delawarr)	50.75	61.20	Freq: 20.0
Eyes, Eyeball, Iris, Nerves, Ciliary Ganglion Branches, (Delawarr)	63.75	65.10	Freq: 20.0
Eyes, Eyeball, Iris, Nerves, From Occulomotor Nerve, (Delawarr)	57.75	68.20	Freq: 20.0
Eyes, Eyeball, Iris, Nerves, Long Ciliary, (Delawarr)	51.00	49.50	Freq: 20.0
Eyes, Eyeball, Iris, Pupil, (Delawarr)	27.00	25.00	Freq: 20.0
Eyes, Eyeball, Iris, Pupil, (Delawarr)	78.60	80.75	Freq: 20.0
Eyes, Eyeball, Iris, Stroma, (Delawarr)	66.20	55.50	Freq: 20.0
Eyes, Eyeball, Lacrimal Apparatus, (Delawarr)	61.20	64.20	Freq: 20.0
Eyes, Eyeball, Lacrimal Apparatus, Lacrimal Ducts, (Delawarr)	00.00	64.00	Freq: 20.0
Eyes, Eyeball, Lacrimal Apparatus, Lacrimal Ducts, (Delawarr)	60.50	59.25	Freq: 20.0
Eyes, Eyeball, Lacrimal Apparatus, Lacrimal Gland, (Delawarr)	43.50	56.60	Freq: 20.0
Eyes, Eyeball, Lacrimal Apparatus, Lacrimal Sac, (Delawarr)	56.40	61.60	Freq: 20.0
Eyes, Eyeball, Lens, (Delawarr)	50.00	38.25	Freq: 20.0
Eyes, Eyeball, Lens, Capsule, (Delawarr)	57.25	56.50	Freq: 20.0
Eyes, Eyeball, Lens, Nucleus, (Delawarr)	56.50	68.50	Freq: 20.0
Eyes, Eyeball, Orbital Cavity, (Delawarr)	35.75	45.75	Freq: 20.0
Eyes, Eyeball, Orbital Cavity, Orbital Muscles, (Delawarr)	50.60	46.25	Freq: 20.0, 240.0
Eyes, Eyeball, Orbital Cavity, Rectus Inferior, (Delawarr)	60.00	27.50	Freq: 20.0
Eyes, Eyeball, Orbital Cavity, Rectus Lateral, (Delawarr)	47.25	48.00	Freq: 20.0
Eyes, Eyeball, Orbital Cavity, Rectus Medial, (Delawarr)	14.00	20.00	Freq: 20.0
Eyes, Eyeball, Orbital Cavity, Rectus Oblique Inferior, (Delawarr)	33.00	43.20	Freq: 20.0
Eyes, Eyeball, Orbital Cavity, Rectus Oblique Superior, (Delawarr)	56.50	56.50	Freq: 20.0
Eyes, Eyeball, Orbital Cavity, Rectus Superior, (Delawarr)	26.75	56.00	Freq: 20.0
Eyes, Eyeball, Retina, (Delawarr)	40.50	54.30	Freq: 20.0
Eyes, Eyeball, Retina, (Delawarr)	48.00	54.00	Freq: 20.0
Eyes, Eyeball, Retina, Arteries, (Delawarr)	48.50	50.25	Freq: 20.0
Eyes, Eyeball, Retina, Ganglionic Layer, (Delawarr)	46.00	50.50	Freq: 20.0
Eyes, Eyeball, Retina, Inner Nuclear Layer, (Delawarr)	42.60	53.80	Freq: 20.0
Eyes, Eyeball, Retina, Inner Plexiform Layer, (Delawarr)	52.40	53.00	Freq: 20.0
Eyes, Eyeball, Retina, Layer Of Rods & Cones, (Delawarr)	33.20	49.50	Freq: 20.0

Patterns of Information	Rate Left	Rate Right	Energetic Processes
Eyes, Eyeball, Retina, Macula Lutea, (Delawarr)	59.00	55.00	Freq: 20.0
Eyes, Eyeball, Retina, Outer Nuclear Layer, (Delawarr)	44.00	59.00	Freq: 20.0
Eyes, Eyeball, Retina, Outer Plexiform Layer, (Delawarr)	40.00	43.10	Freq: 20.0
Eyes, Eyeball, Retina, Stratum Opticum, (Delawarr)	46.00	39.00	Freq: 20.0
Eyes, Eyeball, Vascular Coat, (Delawarr)	31.50	41.50	Freq: 20.0
Eyes, Eyeball, Vascular Coat, Choroid, (Delawarr)	00.00	36.50	Freq: 20.0
Eyes, Eyeball, Vascular Coat, Choroid, (Delawarr)	70.00	55.50	Freq: 20.0
Eyes, Eyeball, Vascular Coat, Choroid, (Delawarr)	25.00	37.00	Freq: 20.0
Eyes, Eyeball, Vascular Coat, Choroid, Basal Lamina, (Delawarr)	48.00	46.00	Freq: 20.0
Eyes, Eyeball, Vascular Coat, Choroid, Capillaries, (Delawarr)	50.25	48.75	Freq: 20.0
Eyes, Eyeball, Vascular Coat, Choroid, Choriocapillary Lamina, (Delawarr)	46.50	45.40	Freq: 20.0
Eyes, Eyeball, Vascular Coat, Choroid, Suprachoroid Lamina, (Delawarr)	64.75	65.30	Freq: 20.0
Eyes, Eyeball, Vascular Coat, Choroid, Vascular Lamina, (Delawarr)	43.00	57.00	Freq: 20.0
Eyes, Eyeball, Vitreous Body, (Delawarr)	32.75	43.20	Freq: 20.0
Eyes, Eyeball, Vitreous Body, Hyaloid Canal, (Delawarr)	56.00	55.25	Freq: 20.0
Eyes, Eyeball, Vitreous Body, Hyaloid Membrane, (Delawarr)	34.00	32.20	Freq: 20.0
Eyes, Eyeball, Vitreous Body, Suspensory Ligament Of Lens, (Delawarr)	61.00	55.20	Freq: 20.0
Fatty Degeneration, (Delawarr)	43.25	33.75	
Female Genital System, Fallopian Tubes, (Delawarr)	53.00	50.50	
Female Genital System, Fallopian Tubes, External, Serous Coat, (Delawarr)	31.75	27.50	
Female Genital System, Fallopian Tubes, Fimbriated End, (Delawarr)	47.50	32.25	
Female Genital System, Fallopian Tubes, Infundibulum, (Delawarr)	26.50	33.50	
Female Genital System, Fallopian Tubes, Internal Mucous Coat, (Delawarr)	50.50	33.50	
Female Genital System, Fallopian Tubes, Left, (Delawarr)	23.00	19.00	
Female Genital System, Fallopian Tubes, Middle Muscular Coat, (Delawarr)	47.50	48.00	
Female Genital System, Fallopian Tubes, Right, (Delawarr)	23.00	17.00	
Female Genital System, Ovaries, (Delawarr)	82.00	65.50	Freq. 1.0, 880.0
Female Genital System, Ovaries, (Delawarr)	03.00	54.00	Freq. 1.0, 880.0
Female Genital System, Ovaries, (Delawarr)	00.00	59.50	Freq. 1.0, 880.0
Female Genital System, Ovaries, Corpus Luteum, (Delawarr)	58.25	45.75	Freq. 1.0, 880.0
Female Genital System, Ovaries, Germinal Epithelium, (Delawarr)	44.10	44.30	Freq. 1.0, 880.0
Female Genital System, Ovaries, Graafian Follicles, (Delawarr)	48.25	46.00	Freq. 1.0, 880.0
Female Genital System, Ovaries, Membrane Granulosa, (Delawarr)	47.00	49.00	Freq. 1.0, 880.0
Female Genital System, Ovaries, Organ, (Delawarr)	00.00	22.00	Freq. 1.0, 880.0
Female Genital System, Ovaries, Ovum, (Delawarr)	41.00	46.50	Freq: 752.0, 1.0, 880.0
Female Genital System, Ovaries, Stroma, (Delawarr)	46.75	23.00	Freq. 1.0, 880.0
Female Genital System, Ovaries, Suspensory Ligament Of, (Delawarr)	11.60	35.30	Freq. 1.0, 880.0
Female Genital System, Ovaries, Tubal Extremity, (Delawarr)	01.50	56.25	Freq. 1.0, 880.0

Patterns of Information	Rate Left	Rate Right	Energetic Processes
Female Genital System, Ovaries, Uterine Extremity, (Delawarr)	15.50	20.00	Freq. 1.0, 880.0
Female Genital System, Uterus, (Delawarr)	82.00	22.00	
Female Genital System, Uterus, (Delawarr)	43.25	39.75	
Female Genital System, Uterus, (Delawarr)	13.50	34.00	
Female Genital System, Uterus, (Delawarr)	10.00	61.00	
Female Genital System, Uterus, (Delawarr)	08.00	23.00	
Female Genital System, Uterus, Arteries, From Internal Iliac Artery, (Delawarr)	61.50	51.00	
Female Genital System, Uterus, Body Of Uterus, (Delawarr)	42.75	40.50	
Female Genital System, Uterus, Body Of Uterus, Fundus, (Delawarr)	33.00	40.25	
Female Genital System, Uterus, Cervix Of Uterus, External Os, Mouth, Opening, (Delawarr)	46.75	43.50	
Female Genital System, Uterus, Cervix Of Uterus, Internal Os, Mouth, Opening, (Delawarr)	40.25	69.75	
Female Genital System, Uterus, Cervix Of Uterus, Mucous Glands, (Delawarr)	82.50	70.00	
Female Genital System, Uterus, Cervix Of Uterus, Neck of, (Delawarr)	00.00	86.00	
Female Genital System, Uterus, Cervix Of Uterus, Neck of, (Delawarr)	00.00	26.00	
Female Genital System, Uterus, Cervix Of Uterus, Neck of, (Delawarr)	27.40	37.40	
Female Genital System, Uterus, Cervix Of Uterus, Parametrium, (Delawarr)	38.75	43.20	Loose Connective Tissue Around the Uterus
Female Genital System, Uterus, External Serous Coat, (Delawarr)	00.00	53.50	
Female Genital System, Uterus, External Serous Coat, (Delawarr)	42.50	32.60	
Female Genital System, Uterus, Internal Mucous Coat, (Delawarr)	47.75	39.50	
Female Genital System, Uterus, Ligaments, Anterior Ligament, (Delawarr)	08.00	05.50	
Female Genital System, Uterus, Ligaments, Posterior Ligament, (Delawarr)	20.00	17.50	
Female Genital System, Uterus, Ligaments, Round Ligaments, (Delawarr)	87.25	64.75	
Female Genital System, Uterus, Ligaments, Two Broad Ligaments, (Delawarr)	15.00	62.00	
Female Genital System, Uterus, Lymph Vessels, (Delawarr)	86.25	10.50	
Female Genital System, Uterus, Muscular Coat, (Delawarr)	39.20	18.25	
Female Genital System, Uterus, Nerves, From Hypogastric Plexus, (Delawarr)	41.25	19.00	
Female Genital System, Uterus, Nerves, From Ovarian Plexus, (Delawarr)	19.00	18.50	
Female Genital System, Uterus, Organ, (Delawarr)	00.00	58.00	
Female Genital System, Uterus, Pelvic Visceromotor Fibers, (Delawarr)	39.40	28.00	
Female Genital System, Uterus, Veins, Ending In Uterine Plexuses, (Delawarr)	62.50	08.25	
Female Genital System, Vagina, (Delawarr)	21.25	54.50	
Female Genital System, Vagina, (Delawarr)	92.00	31.00	
Female Genital System, Vagina, (Delawarr)	92.00	96.00	
Female Genital System, Vagina, (Delawarr)	00.00	81.00	
Female Genital System, Vagina, (Delawarr)	20.00	56.00	
Female Genital System, Vagina, Anterior Wall, (Delawarr)	28.10	06.00	
Female Genital System, Vagina, Arteries, (Delawarr)	51.25	60.25	

Patterns of Information	Rate Left	Rate Right	Energetic Processes
Female Genital System, Vagina, Clitoris, (Delawarr)	77.10	86.10	
Female Genital System, Vagina, Erectile Tissue, (Delawarr)	38.25	17.50	
Female Genital System, Vagina, Labium, (Delawarr)	60.60	36.25	
Female Genital System, Vagina, Lymph Vessels, (Delawarr)	60.75	11.25	
Female Genital System, Vagina, Mucous Membrane, (Delawarr)	11.10	10.00	
Female Genital System, Vagina, Muscular Coat, (Delawarr)	46.50	39.00	
Female Genital System, Vagina, Nerves, From Pelvic Visceromotor Fibers, (Delawarr)	39.75	89.10	
Female Genital System, Vagina, Nerves, From Vaginal Plexuses, (Delawarr)	26.75	15.00	
Female Genital System, Vagina, Perineal Body, (Delawarr)	63.75	46.75	
Female Genital System, Vagina, Posterior Fornix, (Delawarr)	31.00	16.25	
Female Genital System, Vagina, Posterior Wall, (Delawarr)	24.50	15.00	
Female Genital System, Vagina, Veins, (Delawarr)	49.10	35.25	
Female Genital System, Vulva, (Delawarr)	15.00	18.20	
Female Genital System, Vulva, (Delawarr)	00.00	51.50	
Fever, Hay, Asthma, (Delawarr)	08.00	80.00	Freq. 43.3, 20.0, 880.0
Fibroid, Fibrositis, (Delawarr)	33.75	31.00	
Fungus, Bacteria, Infection, Actinomycosis, Lumpy Jaw, (Delawarr)	32.25	37.50	Results in Bony Degeneration of Jaws, Abscesses in Lungs, Intestines, Prolonged Therapy with Penicillin G is effective; Freq: 787.0
Gall Bladder & Biliary Ducts, Biliary Ducts, Bile Duct, (Delawarr)	00.00	23.50	
Gall Bladder & Biliary Ducts, Biliary Ducts, Bile Duct, (Delawarr)	40.50	52.75	
Gall Bladder & Biliary Ducts, Biliary Ducts, Bile Duct, Cystic Duct, (Delawarr)	41.50	51.50	
Gall Bladder & Biliary Ducts, Biliary Ducts, Common Hepatic Duct, (Delawarr)	57.00	44.50	
Gall Bladder & Biliary Ducts, Biliary Ducts, Spiral Valve, (Delawarr)	50.75	47.50	
Gall Bladder & Biliary Ducts, Gall Bladder, (Delawarr)	14.00	64.00	Use For Constipation, Lindane, Various Colds, Intestinal Worms, Pair Color-Green; Emotion-Anger; Sound-Shouting
Gall Bladder & Biliary Ducts, Gall Bladder, Arteries, (Delawarr)	36.50	27.50	
Gall Bladder & Biliary Ducts, Gall Bladder, Arteries, Cystic, (Delawarr)	48.25	45.50	
Gall Bladder & Biliary Ducts, Gall Bladder, Fibromuscular Coat, (Delawarr)	27.00	37.75	
Gall Bladder & Biliary Ducts, Gall Bladder, Helpful Paired with Liver, (Delawarr)	13.00	64.00	Use for Constipation, Lindane, Various Colds, Intestinal Worms, Pair Color-Green; Emotion-Anger; Sound-Shouting
Gall Bladder & Biliary Ducts, Gall Bladder, Lymph Vessels, (Delawarr)	51.50	32.75	
Gall Bladder & Biliary Ducts, Gall Bladder, Mucous Coat, (Delawarr)	24.40	23.80	
Gall Bladder & Biliary Ducts, Gall Bladder, Nerves, (Delawarr)	44.50	47.50	
Gall Bladder & Biliary Ducts, Gall Bladder, Nerves, Celiac Branches, (Delawarr)	24.00	27.20	
Gall Bladder & Biliary Ducts, Gall Bladder, Nerves, of Hepatic Plexus, (Delawarr)	18.25	27.50	
Gall Bladder & Biliary Ducts, Gall Bladder, Nerves, Right Phrenic Fibers, (Delawarr)	20.00	12.00	
Gall Bladder & Biliary Ducts, Gall Bladder, Organ, (Delawarr)	00.00	35.00	Note E#, Balance 13.00-64.00; Freq: 164.3, 330.0
Gall Bladder & Biliary Ducts, Gall Bladder, Serous Coat, (Delawarr)	23.75	14.00	

Patterns of Information	Rate Left	Rate Right	Energetic Processes
Gall Bladder & Biliary Ducts, Gall Bladder, Sphincter, (Delawarr)	30.25	27.00	
Gall Bladder & Biliary Ducts, Gall Bladder, Veins, (Delawarr)	29.25	38.25	
Gall Bladder & Biliary Ducts, Gall Bladder, Veins, Cystic, (Delawarr)	80.00	37.00	
Gastrointestinal Disorders, Appendicitis, Often With Sharp Pain, Relieved by Food, (Delawarr)	44.50	22.00	
Gastrointestinal Disorders, Colitis, Colon, Large Intestine, Inflammation, (Delawarr)	40.00	91.00	
Gastrointestinal Disorders, Stomach, Ulcers & Sores, Pain Increased by Food, (Delawarr)	92.00	85.00	& at Night, Use 5.00-91.00 Prostate
Heart, Arteries, (Delawarr)	36.50	23.60	Freq: 20.0, 162.0
Heart, Arteries, Left Coronary, (Delawarr)	24.00	12.60	Freq: 20.0, 162.0
Heart, Arteries, Left Coronary, Interventricular Branch, (Delawarr)	19.00	12.00	Freq: 20.0, 162.0
Heart, Arteries, Right Coronary, (Delawarr)	11.60	10.10	Freq: 20.0, 162.0
Heart, Arteries, Right Coronary, Interventricular Branch, (Delawarr)	31.75	17.00	Freq: 20.0, 162.0
Heart, Arteries, Right Coronary, Marginal Branch, (Delawarr)	13.75	10.50	Freq: 20.0, 162.0
Heart, Atria Septum, (Delawarr)	11.25	20.50	Freq: 20.0, 162.0
Heart, Atria, (Delawarr)	14.75	16.75	Freq: 20.0, 162.0
Heart, Atria, Left, (Delawarr)	22.75	18.80	Freq: 20.0, 162.0
Heart, Atria, Right, (Delawarr)	23.60	27.20	Freq: 20.0, 162.0
Heart, Atrio-Ventricular Node, (Delawarr)	21.25	16.40	Freq: 20.0, 162.0
Heart, Auricles, (Delawarr)	14.20	14.60	Freq: 20.0, 162.0
Heart, Auricles, Auriculo Ventricular Ring, (Delawarr)	11.75	21.00	Freq: 20.0, 162.0
Heart, Bundle of His, (Delawarr)	13.75	12.00	Freq: 20.0, 162.0
Heart, Endocardium, (Delawarr)	11.50	11.10	Freq: 20.0, 162.0
Heart, Endocardium, Connective Tissue, (Delawarr)	13.50	14.90	Freq: 20.0, 162.0
Heart, Endocardium, Elastic Fibers, (Delawarr)	20.00	12.00	Freq: 20.0, 162.0
Heart, Fossa Ovalis, (Delawarr)	18.90	16.50	Freq: 20.0, 162.0
Heart, Helpful Paired with Small Intestine 21.00-36.00, (Delawarr)	02.00	76.00	Pair Color-Red; Emotion-Happiness; Sound-Laughter; Freq: 20.0, 162.0
Heart, Lymph Vessels, (Delawarr)	22.00	27.40	Freq: 20.0, 162.0
Heart, Lymph Vessels, Deep Plexus, (Delawarr)	18.75	17.75	Freq: 20.0, 162.0
Heart, Lymph Vessels, Superficial Plexus, (Delawarr)	14.20	14.00	Freq: 20.0, 162.0
Heart, Myocardium, (Delawarr)	20.00	19.00	Freq: 20.0, 162.0
Heart, Myocardium, Fibrous Muscular Rings, (Delawarr)	10.20	22.75	Freq: 20.0, 162.0
Heart, Myocardium, Fibrous Muscular Rings, Longitudinal, (Delawarr)	10.75	08.50	Freq: 20.0, 162.0
Heart, Myocardium, Fibrous Muscular Rings, Transverse, (Delawarr)	13.10	04.00	Freq: 20.0, 162.0
Heart, Nerves, (Delawarr)	11.25	20.75	Freq: 20.0, 162.0
Heart, Nerves, Cardiac Plexus, (Delawarr)	12.00 00.00	22.50 06.00	Freq: 20.0, 162.0
Heart, Nerves, Cardiac Plexus, Deep, (Delawarr)	22.50	17.20	Freq: 20.0, 162.0
Heart, Nerves, Cardiac Plexus, Superficial, (Delawarr)	14.50	15.00	Freq: 20.0, 162.0
Heart, Organ, (Delawarr)	00.00	70.00	Amino Acid Carnitine Highest providing fat oxidation which results in more energy; Freq: 20.0, 162.0
Heart, Pericardium, (Delawarr)	04.50	12.50	Freq: 20.0, 162.0
Heart, Pericardium, Arteries, (Delawarr)	13.25	12.75	Freq: 20.0, 162.0
Heart, Pericardium, Fibrous Sac, (Delawarr)	18.75	11.25	
Heart, Pericardium, Lymph Glands, Mediastinal, (Delawarr)	27.75	19.50	Freq: 20.0, 162.0
Heart, Pericardium, Nerves, (Delawarr)	12.40	28.20	Freq: 20.0, 162.0
Heart, Pericardium, Serous Sac, (Delawarr)	03.50	21.75	Freq: 20.0, 162.0
Heart, Purkinje's Fibers, (Delawarr)	26.50	18.00	Freq: 20.0, 162.0

Patterns of Information	Rate Left	Rate Right	Energetic Processes
Heart, Purkinje's System, (Delawarr)	18.20	22.60	Freq: 20.0, 162.0
Heart, Sino-Auricular Node, (Delawarr)	12.00	19.50	Freq: 20.0, 162.0
Heart, Tendon Of The Infundibulum, (Delawarr)	30.75	06.40	Freq: 20.0, 162.0
Heart, Valves, (Delawarr)	16.00	14.00	Freq: 20.0, 162.0
Heart, Valves, Aortic, (Delawarr)	08.00	10.60	Freq: 20.0, 162.0
Heart, Valves, Chordae Tendineae, (Delawarr)	10.40	12.25	Freq: 20.0, 162.0
Heart, Valves, Mitral, (Delawarr)	08.00	10.40	Freq: 20.0, 162.0
Heart, Valves, Papillary Muscles, (Delawarr)	16.40	14.60	Freq: 20.0, 162.0, 240.0
Heart, Valves, Pulmonary, (Delawarr)	09.00	05.60	Freq: 20.0, 162.0
Heart, Valves, Tricuspid, (Delawarr)	11.50	10.00	Freq: 20.0, 162.0
Heart, Veins, (Delawarr)	11.00	08.50	Freq: 20.0, 162.0
Heart, Veins, Anterior Cardiac, (Delawarr)	14.20	16.50	Freq: 20.0, 162.0
Heart, Veins, Coronary Sinus, (Delawarr)	08.25	10.00	Freq: 20.0, 162.0
Heart, Veins, Coronary Sinus, Valve Of Coronary Sinus, (Delawarr)	12.50	08.30	Freq: 20.0, 162.0
Heart, Veins, Great Cardiac, (Delawarr)	12.20	09.30	Freq: 20.0, 162.0
Heart, Veins, Left Marginal, (Delawarr)	13.60	15.30	Freq: 20.0, 162.0
Heart, Veins, Middle Cardiac, (Delawarr)	07.75	06.50	Freq: 20.0, 162.0
Heart, Veins, Oblique Vein Of Left Atrium, (Delawarr)	13.00	21.00	Freq: 20.0, 162.0
Heart, Veins, Posterior Vein Of Left Ventricle, (Delawarr)	04.75	08.50	Freq: 20.0, 162.0
Heart, Veins, Right Marginal, (Delawarr)	10.50	08.25	Freq: 20.0, 162.0
Heart, Veins, Small Cardiac, (Delawarr)	28.00	16.10	Freq: 20.0, 162.0
Heart, Ventricles, (Delawarr)	10.75	16.25	Freq: 20.0, 162.0
Heart, Ventricles, Left, (Delawarr)	13.50	16.50	Freq: 20.0, 162.0
Heart, Ventricles, Right, (Delawarr)	06.60	11.30	Freq: 20.0, 162.0
Heart, Ventricular Septum, (Delawarr)	17.60	18.75	Freq: 20.0, 162.0
Hepatic & Biliary Disorders, Calculus, Biliary, (Delawarr)	07.00	35.00	Abnormal Stony Mass In Bile Tract
Hepatic & Biliary Disorders, Calculus, Biliary, (Delawarr)	39.25	38.00	Abnormal Stony Mass In Bile Tract
Hernia, (Delawarr)	40.00	41.00	Use 59.00-177.00, 31.00-2.00, 5.00-91.00, Then 6.00-81.00, Then 77.00-41.00, Check Staph
Hernia, (Delawarr)	27.25	23.50	Use 59.00-177.00, 31.00-2.00, 5.00-91.00, Then 6.00-81.00, Then 77.00-41.00, Check Staph
Hyperinsulinism, (Delawarr)	28.50	31.75	
Index, Delawarr	25.75	30.00	
Innervation, Impaired, (Delawarr)	32.50	17.00	
Internal Ear, Arteries of the Labyrinth, (Delawarr)	61.00	57.75	Freq: 20.0, 880.0
Internal Ear, Arteries of the Labyrinth, Internal Auditory Artery, (Delawarr)	59.00	54.50	Freq: 20.0, 880.0
Internal Ear, Arteries of the Labyrinth, Stylo-Mastoid Branch Of Auricular Artery, (Delawarr)	76.50	76.75	Freq: 20.0, 880.0
Internal Ear, Auditory Nerve, (Delawarr)	62.20	52.50	Freq: 20.0, 880.0
Internal Ear, Auditory Nerve, Cochlear Nerve, (Delawarr)	33.50	63.75	Freq: 20.0, 880.0
Internal Ear, Auditory Nerve, Spiral Ganglion, (Delawarr)	57.50	57.00	Freq: 20.0, 880.0
Internal Ear, Auditory Nerve, Vestibular Ganglion, (Delawarr)	53.80	55.60	Freq: 20.0, 880.0
Internal Ear, Auditory Nerve, Vestibular Nerve, (Delawarr)	64.60	59.90	Freq: 20.0, 880.0
Internal Ear, Bony Labyrinth, (Delawarr)	68.50	46.00	Freq: 20.0, 880.0
Internal Ear, Bony Labyrinth, Bony Labyrinth, Semicircular Canals, Horizontal Canal, (Delawarr)	40.25	43.25	Freq: 20.0, 880.0
Internal Ear, Bony Labyrinth, Cochlea, (Delawarr)	50.00	43.75	Freq: 20.0, 880.0
Internal Ear, Bony Labyrinth, Cochlea, Basilar Membrane, (Delawarr)	60.00	52.20	Freq: 20.0, 880.0
Internal Ear, Bony Labyrinth, Cochlea, Fenestra Cochleae, (Delawarr)	41.50	54.50	Freq: 20.0, 880.0

Patterns of Information	Rate Left	Rate Right	Energetic Processes
Internal Ear, Bony Labyrinth, Cochlea, Helicotrema, (Delawarr)	40.25	52.80	Freq: 20.0, 880.0
Internal Ear, Bony Labyrinth, Cochlea, Modiolus, (Delawarr)	54.50	44.75	Freq: 20.0, 880.0
Internal Ear, Bony Labyrinth, Cochlea, Osseous Spiral Lamina, (Delawarr)	48.00	44.00	Freq: 20.0, 880.0
Internal Ear, Bony Labyrinth, Cochlea, Scala Tympani, (Delawarr)	35.50	43.00	Freq: 20.0, 880.0
Internal Ear, Bony Labyrinth, Cochlea, Scala Vestibuli, (Delawarr)	31.00	41.50	Freq: 20.0, 880.0
Internal Ear, Bony Labyrinth, Cochlea, Secondary Spiral Lamina, (Delawarr)	40.75	46.60	Freq: 20.0, 880.0
Internal Ear, Bony Labyrinth, Cochlea, Spiral Canal, (Delawarr)	46.25	46.00	Freq: 20.0, 880.0
Internal Ear, Bony Labyrinth, Fibro-Serous Membrane Lining, (Delawarr)	38.50	49.00	Freq: 20.0, 880.0
Internal Ear, Bony Labyrinth, Perilymph, (Delawarr)	43.75	43.00	Freq: 20.0, 880.0
Internal Ear, Bony Labyrinth, Pyramid Of The Vestibule, (Delawarr)	61.50	55.60	Freq: 20.0, 880.0
Internal Ear, Bony Labyrinth, Semicircular Canals, (Delawarr)	48.75	50.50	Freq: 20.0, 880.0
Internal Ear, Bony Labyrinth, Semicircular Canals, Ampullae, (Delawarr)	46.80	35.40	Freq: 20.0, 880.0
Internal Ear, Bony Labyrinth, Semicircular Canals, Crus Commune, (Delawarr)	47.00	47.00	Freq: 20.0, 880.0
Internal Ear, Bony Labyrinth, Semicircular Canals, Posterior Semicircular Canal, (Delawarr)	57.00	52.75	Freq: 20.0, 880.0
Internal Ear, Bony Labyrinth, Semicircular Canals, Superior Semicircular Canal, (Delawarr)	57.00	48.40	Freq: 20.0, 880.0
Internal Ear, Bony Labyrinth, Vestibule, (Delawarr)	45.50	40.40	Freq: 20.0, 880.0
Internal Ear, Bony Labyrinth, Vestibule, Vestibular Crest, (Delawarr)	60.00	50.25	Freq: 20.0, 880.0
Internal Ear, Membraneous Labyrinth, Duct of Cochlea, (Delawarr)	70.50	60.25	Freq: 20.0, 880.0
Internal Ear, Membraneous Labyrinth, Duct Of Cochlea, Basilar Membrane, (Delawarr)	60.50	58.25	Freq: 20.0, 880.0
Internal Ear, Membraneous Labyrinth, Duct Of Cochlea, Vestibular Membrane, (Delawarr)	55.60	42.75	Freq: 20.0, 880.0
Internal Ear, Membraneous Labyrinth, Ductus Utriculosaccularis, (Delawarr)	75.00	54.50	Freq: 20.0, 880.0
Internal Ear, Membraneous Labyrinth, Hair Cells, (Delawarr)	54.25	43.75	Freq: 20.0, 880.0
Internal Ear, Membraneous Labyrinth, Rods Of Corti, Outer, (Delawarr)	58.00	64.50	Freq: 20.0, 880.0
Internal Ear, Membraneous Labyrinth, Saccule, (Delawarr)	83.40	78.75	Freq: 20.0, 880.0
Internal Ear, Membraneous Labyrinth, Semicircular Ducts, (Delawarr)	63.00	52.60	Freq: 20.0, 880.0
Internal Ear, Membraneous Labyrinth, Semicircular Ducts, Ampullae, (Delawarr)	67.75	45.75	Freq: 20.0, 880.0
Internal Ear, Membraneous Labyrinth, Semicircular Ducts, Crista Ampullaris, (Delawarr)	70.10	60.60	Freq: 20.0, 880.0
Internal Ear, Membraneous Labyrinth, Semicircular Ducts, Septum Transversum, (Delawarr)	76.25	73.25	Freq: 20.0, 880.0
Internal Ear, Membranous Labyrinth, (Delawarr)	41.60	36.60	Freq: 20.0, 880.0
Internal Ear, Membranous Labyrinth, Duct Of Cochlea, Lagaena, (Delawarr)	49.75	53.40	Freq: 20.0, 880.0
Internal Ear, Membranous Labyrinth, Duct Of Cochlea, Limbus Laminae Spiralis, (Delawarr)	38.50	28.50	Freq: 20.0, 880.0
Internal Ear, Membranous Labyrinth, Duct Of Cochlea, Stria Vascularis, (Delawarr)	50.75	55.50	Freq: 20.0, 880.0
Internal Ear, Membranous Labyrinth, Endolymph, (Delawarr)	48.00	45.50	Freq: 20.0, 880.0

Patterns of Information	Rate Left	Rate Right	Energetic Processes
Internal Ear, Membranous Labyrinth, Macula Of The Utricle, (Delawarr)	44.20	54.75	Freq: 20.0, 880.0
Internal Ear, Membranous Labyrinth, Macula Of The Utricle, Filaments Auditory Nerve, (Delawarr)	42.25	57.50	Freq: 20.0, 880.0
Internal Ear, Membranous Labyrinth, Rods Of Corti, (Delawarr)	48.50	54.50	Freq: 20.0, 880.0
Internal Ear, Membranous Labyrinth, Rods Of Corti, Inner, (Delawarr)	46.50	81.75	Freq: 20.0, 880.0
Internal Ear, Membranous Labyrinth, Saccule, Canalis Reuniens, (Delawarr)	48.00	70.00	Freq: 20.0, 880.0
Internal Ear, Membranous Labyrinth, Saccule, Filament Auditory Nerve, (Delawarr)	50.00	36.00	Freq: 20.0, 880.0
Internal Ear, Membranous Labyrinth, Saccule, Macula of the Saccule, (Delawarr)	18.20	57.60	Freq: 20.0, 880.0
Internal Ear, Membranous Labyrinth, Saccule, Saccus Endlymphatics, (Delawarr)	49.20	57.25	Freq: 20.0, 880.0
Internal Ear, Membranous Labyrinth, Tectorial Membrane, (Delawarr)	46.50	49.50	Freq: 20.0, 880.0
Internal Ear, Membranous Labyrinth, Utricle, (Delawarr)	40.75	51.20	Freq: 20.0, 880.0
Internal Ear, Spiral Organ, Membranous Labyrinth, (Delawarr)	46.50	71.00	Freq: 20.0, 880.0
Internal Ear, Wax Or Cerumen, (Delawarr)	68.75	75.00	Freq: 20.0, 880.0
Internal Ear, Wax or Cerumen, (Delawarr)	04.00	06.00	Use 40.00-6.00; Freq: 20.0, 880.0
Large Intestine or Colon, Helpful to Pair with Lungs 92.00-83.00, (Delawarr)	08.00	72.00	Pair Color-White; Emotion-Sadness; Sound-Weeping; Freq. 9.0
Large Intestine, Anus, (Delawarr)	00.00	42.50	Freq. 10.0, 880.0
Large Intestine, Anus, (Delawarr)	28.25	24.00	May be Affected by DDE; Freq. 10.0, 880.0
Large Intestine, Ascending Colon, (Delawarr)	35.00	47.25	Freq. 10.0, 880.0, 20.0
Large Intestine, Ascending Colon, Organ, (Delawarr)	00.00	08.00	Note F#; Freq: 176.0, 20.0
Large Intestine, Cecum, (Delawarr)	90.00	66.00	Freq. 10.0, 880.0
Large Intestine, Cecum, (Delawarr)	15.00	35.00	Freq. 10.0, 880.0
Large Intestine, Cecum, Transverse Colon, (Delawarr)	00.00	60.00	Freq. 10.0, 880.0, 20.0
Large Intestine, Descending Colon, (Delawarr)	00.00	62.00	Freq. 10.0, 880.0, 20.0
Large Intestine, Descending Colon, (Delawarr)	00.00	90.00	Freq. 10.0, 880.0, 20.0
Large Intestine, Descending Colon, (Delawarr)	23.75	14.00	Freq. 10.0, 880.0, 20.0
Large Intestine, Descending Colon, (Delawarr)	16.40	15.00	Freq. 10.0, 880.0, 20.0
Large Intestine, Hepatic Flexure, (Delawarr)	00.00	78.00	Freq. 10.0, 880.0
Large Intestine, Ileocecal Valve, (Delawarr)	21.00	69.00	Freq. 10.0, 880.0
Large Intestine, Ileocecal Valve, Frenula, (Delawarr)	10.50	39.50	Freq. 10.0, 880.0
Large Intestine, Lymph Vessels, (Delawarr)	75.50	80.25	Freq. 10.0, 880.0
Large Intestine, Mesentery, (Delawarr)	00.00	00.50	Freq. 10.0, 880.0
Large Intestine, Mesentery, (Delawarr)	33.75	43.50	Freq. 10.0, 880.0
Large Intestine, Omentum, (Delawarr)	29.75	44.00	Freq. 10.0, 880.0
Large Intestine, Pelvic Colon, (Delawarr)	00.00	33.75	Freq. 10.0, 880.0, 20.0
Large Intestine, Pelvic Colon, (Delawarr)	50.75	22.75	Freq. 10.0, 880.0, 20.0
Large Intestine, Pelvic Colon, Pelvic Mesocolon, (Delawarr)	28.25	20.50	Freq. 10.0, 880.0, 20.0
Large Intestine, Rectum, (Delawarr)	85.00	01.00	Freq. 10.0, 880.0
Large Intestine, Rectum, (Delawarr)	50.00	12.00	Freq. 10.0, 880.0
Large Intestine, Rectum, (Delawarr)	12.00	35.00	Freq. 10.0, 880.0
Large Intestine, Rectum, External Sphincter, (Delawarr)	90.25	87.00	Freq. 10.0, 880.0
Large Intestine, Rectum, Internal Sphincter, (Delawarr)	77.25	75.25	Freq. 10.0, 880.0
Large Intestine, Rectum, Organ, (Delawarr)	08.50	90.00	Use 55.00-96.00, Then 40.00-96.00; Freq. 10.0, 880.0
Large Intestine, Rectum, Perineal Flexure, (Delawarr)	77.50	77.00	Freq. 10.0, 880.0
Large Intestine, Rectum, Rectal Ampulla, (Delawarr)	90.50	87.50	Freq. 10.0, 880.0

Patterns of Information	Rate Left	Rate Right	Energetic Processes
Large Intestine, Rectum, Sacral Flexure, (Delawarr)	72.00	67.00	Freq. 10.0, 880.0
Large Intestine, Sigmoid Flexure, (Delawarr)	00.00	12.00	Freq. 10.0, 880.0
Large Intestine, Sigmoid Flexure, (Delawarr)	28.75	42.50	Freq. 10.0, 880.0
Large Intestine, Structure Of, Arteries, Mesenteric Inferior, Branching To Descending, (Delawarr)	85.00	66.00	Colon, Transverse Colon (Left Hypochondriac Region), Pelvic Colon, Rectum; Freq. 10.0, 880.0
Large Intestine, Structure Of, Arteries, Mesenteric Superior, Branching To Cecum,, (Delawarr)	64.75	57.50	Appendix, Ascending Colon, Transverse Colon (Right Hyprochondriac Region); Freq. Freq. 10.0, 880.0
Large Intestine, Structure Of, Arteries, Mesenteric Superior, Rectal Middle, (Delawarr)	25.75	26.50	Freq. 10.0, 880.0
Large Intestine, Structure Of, Arteries, Mesenteric Superior, Rectal Superior, (Delawarr)	25.50	31.00	Freq. 10.0, 880.0
Large Intestine, Structure Of, Arteries, Mesentric Inferior, Rectal Inferior, (Delawarr)	32.75	29.25	Freq. 10.0, 880.0
Large Intestine, Structure Of, Mucous Membrane, (Delawarr)	11.25	14.75	Freq. 10.0, 880.0
Large Intestine, Structure Of, Muscular Coat, (Delawarr)	21.75	28.00	Freq. 10.0, 880.0
Large Intestine, Structure Of, Muscular Coat, Circular Fibers, (Delawarr)	16.40	18.00	Freq. 10.0, 880.0
Large Intestine, Structure Of, Muscular Coat, Longitudinal Fibers, (Delawarr)	20.00	12.00	Freq. 10.0, 880.0
Large Intestine, Structure Of, Serous Coat, (Delawarr)	31.00	18.50	Freq. 10.0, 880.0
Large Intestine, Structure Of, Solitary Lymphatic Nodules, (Delawarr)	18.00	17.50	Freq. 10.0, 880.0
Large Intestine, Transverse Colon, (Delawarr)	41.75	39.25	Freq. 10.0, 880.0, 20.0
Large Intestine, Transverse Colon, Left Colic Flexure, (Delawarr)	26.20	10.40	Freq. 10.0, 880.0, 20.0
Large Intestine, Transverse Colon, Phrenicocolic Ligament, (Delawarr)	13.00	06.00	Freq. 10.0, 880.0, 20.0
Large Intestine, Transverse Colon, Transverse Mesocolon, (Delawarr)	17.90	16.20	Freq. 10.0, 880.0, 20.0
Large Intestine, Veins, Hemorrhoidal Venous Plexus, (Delawarr)	83.75	56.00	Freq. 10.0, 880.0
Large Intestine, Veins, Nerves, Parasymphathetics, Pelvic Splanchnic Nerves, (Delawarr)	79.50	68.75	Viscera in Mesenteric Inferior are Served by; Freq. 10.0, 880.0
Large Intestine, Veins, Nerves, Parasymphathetics, Vagus Branch of Large Intestine, (Delawarr)	80.25	62.00	Viscera in Mesenteric Superior are served by; Freq. 10.0, 880.0
Large Intestine, Veins, Nerves, Sympathetic, Superior Mesenteric Ganglion, (Delawarr)	49.75	59.50	Viscera in Mesenteric Superior are served by; Freq. 10.0, 880.0
Large Intestine, Veins, Nerves, Sympathetics, Celiac Ganglion, (Delawarr)	70.50	77.75	Viscera in Mesenteric Superior are served by; Freq. 10.0, 880.0
Large Intestine, Veins, Nerves, Sympathetics, Hypogastric Plexus, (Delawarr)	77.25	62.75	Viscera in Mesenteric Inferior are Served by; Freq. 10.0, 880.0
Large Intestine, Veins, Nerves, Sympathetics, Part Of The Lumbar Plexus, (Delawarr)	56.00	48.25	Viscera in Mesenteric Inferior are Served by; Freq. 10.0, 880.0
Large Intestine, Vermiform Appendix, (Delawarr)	03.00	08.00	Balance For Intestinal Gas (Flatulence), Typhoid Fever, Exanthematous Viral & Respiratory Viral Diseases, Malaria, Thyroid Imbalances, Eczema; Freq. 10.0, 880.0
Large Intestine, Vermiform Appendix, (Delawarr)	40.00	46.00	Balance For Intestinal Gas (Flatulence), Typhoid Fever, Exanthematous Viral & Respiratory Viral Diseases, Malaria, Thyroid Imbalances, Eczema; Freq. 10.0, 880.0
Large Intestine, Vermiform Appendix, Circular Muscular Fibers, (Delawarr)	13.50	07.20	Freq. 10.0, 880.0
Large Intestine, Vermiform Appendix, Longitudinal Muscular Fibers, (Delawarr)	23.60	44.60	Freq. 10.0, 880.0
Large Intestine, Vermiform Appendix, Muscous Membrane, (Delawarr)	17.25	20.75	Freq. 10.0, 880.0
Large Intestine, Vermiform Appendix, Organ, (Delawarr)	00.00	66.00	Balance For Intestinal Gas (Flatulence), Typhoid Fever, Exanthematous Viral & Respiratory Viral Diseases, Malaria, Thyroid Imbalances, Eczema; Freq. 10.0, 880.0
Large Intestine, Vermiform Appendix, Serous Coat, (Delawarr)	44.50	41.00	Freq. 10.0, 880.0

Patterns of Information	Rate Left	Rate Right	Energetic Processes
Large Intestine, Vermiform Appendix, Sub Mucous Coat, (Delawarr)	23.60	19.20	Freq. 10.0, 880.0
Lesion, (Delawarr)	24.50	05.00	
Liver, Bile Pigment, (Delawarr)	42.50	40.25	
Liver, Filter Factor, (Delawarr)	07.50	29.00	To Heal, Regenerate, Make Better
Liver, Hepatic Artery, (Delawarr)	45.50	40.75	
Liver, Hepatic Artery, Conveying Arterial Blood To, Bile Ductules, (Delawarr)	29.00	33.00	
Liver, Hepatic Artery, Conveying Arterial Blood To, Connective Tissue, (Delawarr)	33.00	46.10	
Liver, Hepatic Artery, Conveying Arterial Blood To, Interlobular Plexuses, (Delawarr)	07.25	20.50	
Liver, Hepatic Artery, Conveying Arterial Blood To, Walls Of Portal Vein, (Delawarr)	33.50	34.00	
Liver, Hepatic Artery, Conveying Oxygenated Blood To, Bile Capillaries, (Delawarr)	25.50	29.00	
Liver, Hepatic Artery, Conveying Oxygenated Blood To, Bile Ductules, (Delawarr)	33.25	30.25	
Liver, Hepatic Artery, Conveying Oxygenated Blood To, Liver Cells, (Delawarr)	21.25	35.50	
Liver, Hepatic Artery, Conveying Oxygenated Blood To, Liver Cells, (Delawarr)	41.00	39.75	
Liver, Hepatic Artery, Conveying Oxygenated Blood To, Sinusoids, (Delawarr)	30.75	36.25	
Liver, Left Lobe, (Delawarr)	00.00	45.00	
Liver, Left Lobe, (Delawarr)	45.00	29.00	Use For Allergy, Poisons, Cancer, Diabetes, Edema, Phlegm, Insect Stings, Undulant Fever, Protein Digestion, Constipation, Mumps, Leukemia
Liver, Left Lobe, (Delawarr)	43.20	28.00	
Liver, Left Lobe, Coronary Ligament, (Delawarr)	29.00	38.00	
Liver, Left Lobe, Falciform Ligament, (Delawarr)	29.00	18.00	
Liver, Left Lobe, Left Triangular Ligament, (Delawarr)	36.00	28.00	
Liver, Left Lobe, Right Triangular Ligament, (Delawarr)	42.50	48.25	
Liver, Liver Factor, (Delawarr)	12.00	29.00	To Build, Increase, Enhance
Liver, Lymph Vessels, Deep, (Delawarr)	40.00	38.50	
Liver, Lymph Vessels, Hepatic Artery, (Delawarr)	45.00	29.25	
Liver, Lymph Vessels, Superficial, (Delawarr)	51.75	49.50	
Liver, Nerves, (Delawarr)	52.00	36.00	
Liver, Nerves, From Celiac Plexus, (Delawarr)	44.25	39.00	
Liver, Nerves, From Right Phrenic, (Delawarr)	33.00	16.00	
Liver, Nerves, From Vagus Nerves, (Delawarr)	30.75	34.50	
Liver, Nerves, Sympathetic, (Delawarr)	45.50	25.00	
Liver, Organ, (Delawarr)	00.00	29.00	Note Eb#, Ab#. Freq. 317.83, 198.0
Liver, Posterior Surface, Groove for Inferior Vena Cava, (Delawarr)	46.75	41.50	
Liver, Right Lobe, (Delawarr)	72.00	29.00	
Liver, Right Lobe, (Delawarr)	27.75	30.00	
Liver, Right Lobe, (Delawarr)	18.00	29.00	
Liver, Right Lobe, Caudate Lobe, (Delawarr)	47.50	37.00	
Liver, Right Lobe, Quadrate Lobe, (Delawarr)	40.25	45.00	
Liver, Right Lobe, Quadrate Lobe, (Delawarr)	40.25	35.00	
Liver, Right Lobe, Spigelius's Lobe, (Delawarr)	36.75	48.50	
Liver, Surfaces, Anterior Surface, (Delawarr)	30.75	43.25	
Liver, Surfaces, Fissure For Ligamentum Venosum, (Delawarr)	40.30	39.00	
Liver, Surfaces, Inferior or Visceral Surface, (Delawarr)	40.25	20.75	
Liver, Surfaces, Inferior or Visceral Surface, Colic Impression, (Delawarr)	27.00	27.50	
Liver, Surfaces, Inferior or Visceral Surface, Duodenal Impression, (Delawarr)	34.75	24.50	

Patterns of Information	Rate Left	Rate Right	Energetic Processes
Liver, Surfaces, Inferior or Visceral Surface, Gastric Impression, (Delawarr)	35.00	26.75	
Liver, Surfaces, Inferior or Visceral Surface, Porta Impression, (Delawarr)	26.75	24.00	
Liver, Surfaces, Inferior or Visceral Surface, Renal Impression, (Delawarr)	35.00	31.50	
Liver, Surfaces, Posterior Surface, (Delawarr)	34.00	38.75	
Liver, Surfaces, Posterior Surface, Suprarenal (Adrenal) Impression, (Delawarr)	54.25	38.50	
Liver, Surfaces, Right Surface, (Delawarr)	36.75	52.00	
Liver, Surfaces, Superior Surface, (Delawarr)	34.25	40.75	
Liver, Veins, (Delawarr)	20.00	23.75	
Liver, Veins, Hepatic, (Delawarr)	31.00	41.50	3 vessels returning blood from the liver and discharging into the inferior vena cava
Liver, Veins, Portal, (Delawarr)	30.20	36.25	Formed By The Veins Of The Splanchnic Area Conveying Its Blood Into The Liver
Lymphatic System, (Delawarr)	35.00	37.50	Freq. 10.0, 880.0
Lymphatic System, Endothelium Lining to Glands, (Delawarr)	38.75	43.60	Freq. 10.0, 880.0
Lymphatic System, Lymph Glands, (Delawarr)	00.00	17.00	Freq. 10.0, 880.0
Lymphatic System, Lymph Glands, (Delawarr)	03.00	100.00	Freq. 10.0, 880.0
Lymphatic System, Lymph Glands, (Delawarr)	43.00	45.25	Freq. 10.0, 880.0
Lymphatic System, Lymph Glands, Auricularr, Draining Area Of Ear, (Delawarr)	32.25	38.75	Freq: 20.0, 880.0, 10.0
Lymphatic System, Lymph Glands, Axillary, Draining Upper Limbs & Breasts, (Delawarr)	41.00	35.75	Freq. 10.0, 880.0
Lymphatic System, Lymph Glands, Bronchial, Draining The Lungs, (Delawarr)	37.20	41.20	Freq. 9.0, 10.0, 880.0
Lymphatic System, Lymph Glands, Celiac, Draining Stomach, Spleen, Pancreas, Liver, (Delawarr)	40.50	35.20	Freq. 10.0, 880.0
Lymphatic System, Lymph Glands, Deep Cervical, Draining The Tongue, Lower Pharynx,, (Delawarr)	39.25	29.50	Larynx &Thyroid; Freq. 10.0, 880.0
Lymphatic System, Lymph Glands, Deep Inguinal, Draining Deep Tissues Of Lower Limbs, (Delawarr)	45.00	29.00	Use For Parasites, Hookworm, Pinworm, Arbovirus, Arenavirus, Encephalitis, Bronchitis, Pulmonary Disorders, Fright, Emphysema, Emotional Shock; Freq. 10.0, 880.0
Lymphatic System, Lymph Glands, Epitrochlear, Draining The Hands & Forearms, (Delawarr)	37.75	37.00	Freq. 10.0, 880.0
Lymphatic System, Lymph Glands, External Iliac, Draining The Groin, (Delawarr)	35.00	28.80	Freq. 10.0, 880.0
Lymphatic System, Lymph Glands, Intercostal, Draining The Chest Wall, (Delawarr)	36.25	27.20	Freq. 10.0, 880.0
Lymphatic System, Lymph Glands, Internal Iliac, Draining The Pelvic Viscera, (Delawarr)	59.50	44.00	Freq. 10.0, 880.0
Lymphatic System, Lymph Glands, Lingual, Draining Deeper Tissues Of The Neck, (Delawarr)	76.00	35.40	Freq. 10.0, 880.0
Lymphatic System, Lymph Glands, Lumbar, Draining Kidneys, Adrenals, Ovaries,, (Delawarr)	67.00	38.50	& Testicles. Freq. 1.0, 880.0, 10.0
Lymphatic System, Lymph Glands, Lymph Nodes, (Delawarr)	36.00	17.00	Help Battle, Virus & Other Abnormal Cellular Tissue, Check congestion, Inflammation, Toxins & Bacteria
Lymphatic System, Lymph Glands, Mediastinal, Draining Heart, Pericardium & Thymus, (Delawarr)	48.80	29.80	Freq. 10.0, 880.0
Lymphatic System, Lymph Glands, Mesenteric, Draining Intestines, (Delawarr)	72.75	46.50	Freq. 10.0, 880.0
Lymphatic System, Lymph Glands, Popliteal, Draining The Legs, Body-Back of Thighs, (Delawarr)	38.75	35.00	Freq. 10.0, 880.0
Lymphatic System, Lymph Glands, Sternal, Draining Front of Chest Wall & The, (Delawarr)	47.00	35.50	Abdominal Wall; Freq. 10.0, 880.0
Lymphatic System, Lymph Glands, Sub-Maxillary, Draining Face & Floor of Mouth, (Delawarr)	47.00	68.60	Freq. 10.0, 880.0
Lymphatic System, Lymph Glands, Sub-occipital, Draining Back Of The Scalp, (Delawarr)	31.75	38.50	Freq. 10.0, 880.0

Patterns of Information	Rate Left	Rate Right	Energetic Processes
Lymphatic System, Lymph Glands, Superficial Cervical, Draining External Ear & Neck, (Delawarr)	44.25	29.20	Freq. 10.0, 880.0
Lymphatic System, Lymph Glands, Superficial Inguinal, Draining Buttocks, Anus,, (Delawarr)	57.25	38.50	Perineum, External Genitals & Nearly All The Superficial Lymph Of Lower Limbs; Freq. 10.0, 880.0
Lymphatic System, Lymph Vessels, External Coat, (Delawarr)	33.25	34.20	Freq. 10.0, 880.0
Lymphatic System, Lymph Vessels, Internal Coat, (Delawarr)	37.75	36.50	Freq. 10.0, 880.0
Lymphatic System, Lymph Vessels, Middle Coat, (Delawarr)	25.00	26.00	Freq. 10.0, 880.0
Lymphatic System, Lymph, Fluid, (Delawarr)	38.25	37.50	Freq. 10.0, 880.0
Lymphatic System, Lymph, Lymphocytes, Lymphoid Activity, Cloning Sites for New, (Delawarr)	45.00	37.00	Immune Cells, Blood, Red Cells, White Pulp; Freq. 10.0, 880.0
Lymphatic System, Lymph, Pericardial Lymph Fluid, (Delawarr)	39.50	43.00	Freq. 10.0, 880.0
Lymphatic System, Lymph, Vessels, (Delawarr)	34.75	27.60	Freq. 10.0, 880.0
Lymphatic System, Right Lymphatic Duct, Tributaries, Right Bronchomediastinal Trunk, (Delawarr)	33.75	27.25	Freq. 10.0, 880.0
Lymphatic System, Right Lymphatic Duct, Tributaries, Right Jugular Trunk, (Delawarr)	30.50	36.00	Freq. 10.0, 880.0
Lymphatic System, Right Lymphatic Duct, Tributaries, Right Suclavian Trunk, (Delawarr)	24.00	38.25	Freq. 10.0, 880.0
Lymphatic System, Thoracic Duct, (Delawarr)	26.00	33.00	Freq. 10.0, 880.0
Lymphatic System, Thoracic Duct, Cisterna Chyli, (Delawarr)	29.75	29.50	Freq. 10.0, 880.0
Lymphatic System, Thoracic Duct, Tributaries, Left Bronchomediastinal Trunk, (Delawarr)	38.50	37.00	Freq. 10.0, 880.0
Lymphatic System, Thoracic Duct, Tributaries, Left Jugular Trunk, (Delawarr)	31.25	25.50	Freq. 10.0, 880.0
Lymphatic System, Thoracic Duct, Tributaries, Left Subclavian Trunk, (Delawarr)	38.25	37.75	Freq. 10.0, 880.0
Lymphatics Disorder, Lymph Glands, Lymphadenoma, (Delawarr)	25.25	25.50	Freq. 10.0, 880.0
Male Genital System, Penis, (Delawarr)	06.00	88.00	
Male Genital System, Penis, (Delawarr)	00.00	76.50	
Male Genital System, Penis, Arteries, (Delawarr)	64.25	58.00	
Male Genital System, Penis, Arteries, Helicine Arteries, (Delawarr)	77.50	64.00	
Male Genital System, Penis, Body, (Delawarr)	64.00	54.25	
Male Genital System, Penis, Body, Corpora Cavernosum Penis, (Delawarr)	82.00	71.50	
Male Genital System, Penis, Body, Corpus Spongiosum Penis, (Delawarr)	53.25	62.00	
Male Genital System, Penis, Body, Glans Penis, (Delawarr)	75.50	89.00	
Male Genital System, Penis, Body, Neck Of Penis, (Delawarr)	88.75	46.50	
Male Genital System, Penis, Bulb of Penis, (Delawarr)	35.25	52.25	
Male Genital System, Penis, Nerves, (Delawarr)	68.25	59.00	
Male Genital System, Penis, Nerves, From 2nd Sacral, (Delawarr)	44.75	52.20	
Male Genital System, Penis, Nerves, From 3rd Sacral, (Delawarr)	75.50	86.50	
Male Genital System, Penis, Nerves, From 4th Sacral, (Delawarr)	52.50	52.75	
Male Genital System, Penis, Nerves, Pudenal Branch, (Delawarr)	60.75	35.75	
Male Genital System, Penis, Root, (Delawarr)	71.25	54.25	
Male Genital System, Penis, Root, Crus, (Delawarr)	33.60	45.00	
Male Genital System, Penis, Veins, (Delawarr)	82.00	69.50	

Patterns of Information	Rate Left	Rate Right	Energetic Processes
Male Genital System, Prostate Gland, (Delawarr)	13.50	26.00	Surrounds the Neck of the Bladder & Urethra In Males, It is Part Glandular & part Muscular, Gland Secretes a Thin Opalescent, Slightly Alkaline Fluid Which Is Part of Semen; Freq: 20.0, 2000.0
Male Genital System, Prostate Gland, (Delawarr)	13.00	26.00	Surrounds the Neck of the Bladder & Urethra In Males, It is Part Glandular & part Muscular, Gland Secretes a Thin Opalescent, Slightly Alkaline Fluid Which Is Part of Semen; Freq: 20.0, 2000.0
Male Genital System, Prostate Gland, (Delawarr)	05.00	91.00	Surrounds the Neck of the Bladder & Urethra In Males, It is Part Glandular & part Muscular, Gland Secretes a Thin Opalescent, Slightly Alkaline Fluid Which Is Part of Semen; Freq: 20.0, 2000.0
Male Genital System, Prostate Gland, (Delawarr)	58.00	26.00	Surrounds the Neck of the Bladder & Urethra In Males, It is Part Glandular & part Muscular, Gland Secretes a Thin Opalescent, Slightly Alkaline Fluid Which Is Part of Semen; Freq: 20.0, 2000.0
Male Genital System, Prostate Gland, (Delawarr)	52.75	82.50	Surrounds the Neck of the Bladder & Urethra In Males, It is Part Glandular & part Muscular, Gland Secretes a Thin Opalescent, Slightly Alkaline Fluid Which Is Part of Semen; Freq: 20.0, 2000.0
Male Genital System, Prostate Gland, Arteries, (Delawarr)	57.50	51.75	Freq: 20.0, 2000.0
Male Genital System, Prostate Gland, Glandular Substance, (Delawarr)	73.75	63.75	A Secretion Which is thin, opalescent, slightly alkaline fluid which forms part of the semen; Freq: Freq: 20.0, 2000.0
Male Genital System, Prostate Gland, Glandular Tissue, (Delawarr)	85.00	56.00	Freq: 20.0, 2000.0
Male Genital System, Prostate Gland, Interstitial Tissue, (Delawarr)	83.25	61.50	Freq: 20.0, 2000.0
Male Genital System, Prostate Gland, Lateral Lobes, (Delawarr)	84.25	61.00	Freq: 20.0, 2000.0
Male Genital System, Prostate Gland, Lymph Vessels, (Delawarr)	71.75	57.75	Freq: 20.0, 2000.0
Male Genital System, Prostate Gland, Muscular Tissue, (Delawarr)	62.75	69.00	Freq: 20.0, 2000.0
Male Genital System, Prostate Gland, Nerves, (Delawarr)	19.50	95.00	Freq: 20.0, 2000.0
Male Genital System, Prostate Gland, Organ, (Delawarr)	00.00	26.00	Surrounds the neck of the bladder and urethra in males, it is part glandular and part muscular, gland secretes a thin opalescent, slightly alkaline fluid which is part of the semen. Freq. 20.0, 2000.0
Male Genital System, Prostate Gland, Veins, (Delawarr)	13.75	21.50	Freq: 20.0, 2000.0
Male Genital System, Scrotum, (Delawarr)	02.00	98.00	
Male Genital System, Scrotum, (Delawarr)	25.60	25.75	
Male Genital System, Scrotum, Dartos Muscle, (Delawarr)	36.60	33.00	Freq: 20.0, 240.0
Male Genital System, Scrotum, Skin, (Delawarr)	27.50	35.50	
Male Genital System, Seminal Vesicles, (Delawarr)	75.50	63.50	
Male Genital System, Seminal Vesicles, and Ejaculatory Ducts, (Delawarr)	81.50	48.60	
Male Genital System, Seminal Vesicles, Internal Mucous Coat, (Delawarr)	34.25	37.50	
Male Genital System, Seminal Vesicles, Muscular Coat, (Delawarr)	68.25	53.50	
Male Genital System, Seminal Vesicles, Nerves Derived From Pelvic Plexuses, (Delawarr)	40.25	37.00	
Male Genital System, Seminal Vesicles, Outer Fibrous Layer, (Delawarr)	22.25	48.50	
Male Genital System, Spermatic Cord, (Delawarr)	00.00	86.00	High concentrations of amino acid carnitine, which provides energy for sperms motility
Male Genital System, Spermatic Cord, (Delawarr)	38.10	36.50	
Male Genital System, Spermatic Cord, Arteries, (Delawarr)	56.50	50.25	
Male Genital System, Spermatic Cord, Cremasteric Fascia, (Delawarr)	30.75	47.60	
Male Genital System, Spermatic Cord, External Spermatic Fascia, (Delawarr)	57.25	55.00	
Male Genital System, Spermatic Cord, Internal Spermatic Fascia, (Delawarr)	31.75	34.75	

Patterns of Information	Rate Left	Rate Right	Energetic Processes
Male Genital System, Spermatic Cord, Nerves, (Delawarr)	54.25	44.50	
Male Genital System, Testes, (Delawarr)	37.25	35.50	Freq. 1.0, 880.0
Male Genital System, Testes, (Delawarr)	37.00	94.00	Freq. 1.0, 880.0
Male Genital System, Testes, Epididymis, (Delawarr)	34.40	33.50	Freq. 1.0, 880.0
Male Genital System, Testes, Lymph Vessels, (Delawarr)	32.75	31.50	Freq. 1.0, 880.0
Male Genital System, Testes, Lymph Vessels, Pre-aortic Lymph Glands, (Delawarr)	27.25	14.75	Freq. 1.0, 880.0
Male Genital System, Testes, Nerves, (Delawarr)	27.50	27.60	Freq. 1.0, 880.0
Male Genital System, Testes, Nerves, From 10th Thoracic Segment, (Delawarr)	73.00	46.50	Freq. 1.0, 880.0
Male Genital System, Testes, Nerves, From Aortic Plexus, (Delawarr)	68.25	57.50	Freq. 1.0, 880.0
Male Genital System, Testes, Nerves, From Renal Plexus, (Delawarr)	72.25	30.50	Freq. 1.0, 880.0
Male Genital System, Testes, Organ, (Delawarr)	00.00	22.00	Freq. 1.0, 880.0
Male Genital System, Testes, Testicular Artery, (Delawarr)	33.25	26.75	Freq. 1.0, 880.0
Male Genital System, Testes, Testicular Veins, (Delawarr)	26.75	12.75	Freq. 1.0, 880.0
Male Genital System, Testes, Testicular Veins, Pampiniform Plexus, (Delawarr)	29.50	27.25	Freq. 1.0, 880.0
Male Genital System, Testes, Tunica Albuginea, (Delawarr)	26.75	20.25	Freq. 1.0, 880.0
Male Genital System, Testes, Tunica Vaginalis, (Delawarr)	30.25	28.00	Freq. 1.0, 880.0
Male Genital System, Testes, Tunica Vasculosa, (Delawarr)	28.50	18.50	Freq. 1.0, 880.0
Male Genital System, Vas Deferens, (Delawarr)	36.60	59.75	
Male Genital System, Vas Deferens, Ductili Aberrantes, (Delawarr)	77.25	48.75	
Metabolism, Imbalance, (Delawarr)	08.25	29.25	
Middle Ear, Auditory Ossicles, (Delawarr)	49.00	62.75	Freq: 20.0, 880.0
Middle Ear, Auditory Ossicles, Articulations Of Ossicles, (Delawarr)	48.80	49.00	Freq: 20.0, 880.0
Middle Ear, Auditory Ossicles, Incus, (Delawarr)	41.50	49.60	Freq: 20.0, 880.0
Middle Ear, Auditory Ossicles, Incus, Ligaments Of Incus, (Delawarr)	34.00	32.00	Freq: 20.0, 880.0
Middle Ear, Auditory Ossicles, Malleus Ligaments Of Malleus, (Delawarr)	62.10	47.20	Freq: 20.0, 880.0
Middle Ear, Auditory Ossicles, Malleus, (Delawarr)	55.50	61.75	Freq: 20.0, 880.0
Middle Ear, Auditory Ossicles, Stapes, (Delawarr)	57.00	44.40	Freq: 20.0, 880.0
Middle Ear, Auditory Ossicles, Stapes, Ligaments Of Stapes, (Delawarr)	46.75	49.00	Freq: 20.0, 880.0
Middle Ear, Pharyngotympanic Tube, Auditory Tube, (Delawarr)	45.75	48.75	Freq: 20.0, 880.0
Middle Ear, Pharyngotympanic Tube, Auditory Tube, Bony Part, (Delawarr)	48.20	60.25	Freq: 20.0, 880.0
Middle Ear, Pharyngotympanic Tube, Auditory Tube, Cartilagenous Part, (Delawarr)	59.00	55.20	Freq: 20.0, 880.0
Middle Ear, Pharyngotympanic Tube, Auditory Tube, Mucous Membrane, (Delawarr)	51.40	50.20	Freq: 20.0, 880.0
Middle Ear, Pharyngotympanic Tube, Auditory Tube, Tube Tonsil, (Delawarr)	51.50	55.25	Freq: 20.0, 880.0
Mouth, Nose & Throat, Body-Mouth, (Delawarr)	11.00	81.00	
Mouth, Nose & Throat, Body-Nose, (Delawarr)	00.00	23.50	
Mouth, Nose & Throat, Esophagus, (Delawarr)	17.25	29.25	
Mouth, Nose & Throat, Esophagus, (Delawarr)	20.00	43.00	
Mouth, Nose & Throat, Esophagus, Mucous Membrane, (Delawarr)	19.40	21.25	

Patterns of Information	Rate Left	Rate Right	Energetic Processes
Mouth, Nose & Throat, Esophagus, Muscle Fibers, Circular, (Delawarr)	20.00	24.00	Freq: 20.0, 240.0
Mouth, Nose & Throat, Esophagus, Muscle, Fibers, (Delawarr)	18.75	24.50	Freq: 20.0, 240.0
Mouth, Nose & Throat, Esophagus, Muscule Fibers, Longitudinal, (Delawarr)	13.50	19.00	
Mouth, Nose & Throat, Esophagus, Stratified Epithelium, (Delawarr)	20.00	21.00	
Mouth, Nose & Throat, Esophagus, Sub Mucous Coat, (Delawarr)	23.75	24.00	
Mouth, Nose & Throat, Mouth, Arteries, (Delawarr)	21.25	19.50	
Mouth, Nose & Throat, Mouth, Gums, (Delawarr)	03.00	25.00	
Mouth, Nose & Throat, Mouth, Gums, Toothpaste Program (11) Heal/Prevent, (Delawarr)	51.00	25.00	
Mouth, Nose & Throat, Mouth, Lips, Bottom, (Delawarr)	32.75	34.75	
Mouth, Nose & Throat, Mouth, Lips, Top, (Delawarr)	31.50	34.50	
Mouth, Nose & Throat, Mouth, Lymph Vessels, (Delawarr)	76.40	56.50	
Mouth, Nose & Throat, Mouth, Nerves, (Delawarr)	41.00	42.00	
Mouth, Nose & Throat, Mouth, Teeth, (Delawarr)	04.00	99.00	
Mouth, Nose & Throat, Mouth, Teeth, Toothpaste Program (12) Whiten/Vitalize, (Delawarr)	59.00	20.00	
Mouth, Nose & Throat, Mouth, Tongue, (Delawarr)	85.00	43.00	
Mouth, Nose & Throat, Mouth, Tongue, Circumvallate Papillae, (Delawarr)	00.40	24.50	
Mouth, Nose & Throat, Mouth, Tongue, Extrinsic Muscles, (Delawarr)	11.75	46.25	Freq: 20.0, 240.0
Mouth, Nose & Throat, Mouth, Tongue, Filliform Papillae, (Delawarr)	41.50	77.00	
Mouth, Nose & Throat, Mouth, Tongue, Fungiform Papillae, (Delawarr)	10.50	07.80	
Mouth, Nose & Throat, Mouth, Tongue, Mucous Membrane, (Delawarr)	12.50	59.00	
Mouth, Nose & Throat, Mouth, Tongue, Pharynx, Eustachian Tubes, Taste Buds, (Delawarr)	00.00	43.50	
Mouth, Nose & Throat, Mouth, Tongue, Taste Buds, (Delawarr)	32.75	28.75	
Mouth, Nose & Throat, Mouth, Tongue, Taste Buds, Bitterness, (Delawarr)	30.50	25.00	
Mouth, Nose & Throat, Mouth, Tongue, Taste Buds, Sourness, (Delawarr)	32.25	36.50	
Mouth, Nose & Throat, Mouth, Tongue, Taste Buds, Sweetness, (Delawarr)	15.50	29.75	
Mouth, Nose & Throat, Mouth, Veins, (Delawarr)	42.00	49.25	
Mouth, Nose & Throat, Nose, (Delawarr)	19.00	96.00	Freq: 20.0, 880.0
Mouth, Nose & Throat, Nose, (Delawarr)	51.00	53.00	Freq: 20.0, 880.0
Mouth, Nose & Throat, Nose, Anterior Nares, (Delawarr)	14.75	21.25	Freq: 20.0, 880.0
Mouth, Nose & Throat, Nose, Lower Nasal Cartilage, (Delawarr)	21.00	53.00	Freq: 20.0, 880.0
Mouth, Nose & Throat, Nose, Nasal Cavity, (Delawarr)	20.75	17.25	Freq: 20.0, 880.0
Mouth, Nose & Throat, Nose, Nasal Cavity, (Delawarr)	32.00	48.00	Freq: 20.0, 880.0
Mouth, Nose & Throat, Nose, Nerves For Muscles, Facial Nerve, (Delawarr)	45.75	51.00	Freq: 20.0, 880.0, 240.0
Mouth, Nose & Throat, Nose, Nerves For Skin, (Delawarr)	40.75	50.60	Freq: 20.0, 880.0
Mouth, Nose & Throat, Nose, Olfactory Region, (Delawarr)	16.25	17.00	Freq: 20.0, 880.0

Patterns of Information	Rate Left	Rate Right	Energetic Processes
Mouth, Nose & Throat, Nose, Posterior Nares, (Delawarr)	17.00	16.75	Freq: 20.0, 880.0
Mouth, Nose & Throat, Nose, Respiratory Region, (Delawarr)	20.00	20.75	Freq: 20.0, 880.0
Mouth, Nose & Throat, Nose, Respiratory Region, Arteries, (Delawarr)	22.60	23.50	Freq: 20.0, 162.0
Mouth, Nose & Throat, Nose, Respiratory Region, Lateral Wall, (Delawarr)	20.50	27.25	Freq: 20.0, 880.0
Mouth, Nose & Throat, Nose, Respiratory Region, Nasal Muscous Membrane, (Delawarr)	20.60	22.75	Freq: 20.0, 880.0
Mouth, Nose & Throat, Nose, Respiratory Region, Nasal Septum, (Delawarr)	23.40	20.00	Freq: 20.0, 880.0
Mouth, Nose & Throat, Nose, Respiratory Region, Nasolacrimal Duct, (Delawarr)	20.75	19.50	Freq: 20.0, 880.0
Mouth, Nose & Throat, Nose, Respiratory Region, Nerves, (Delawarr)	11.50	12.50	Freq: 20.0, 880.0
Mouth, Nose & Throat, Nose, Respiratory Region, Veins, (Delawarr)	24.00	21.50	Freq: 20.0, 880.0
Mouth, Nose & Throat, Nose, Upper Nasal Cartilage, (Delawarr)	09.75	16.50	Freq: 20.0, 880.0
Mouth, Nose & Throat, Pharynx, (Delawarr)	21.75	33.00	
Mouth, Nose & Throat, Pharynx, (Delawarr)	24.00	07.00	
Mouth, Nose & Throat, Pharynx, Adenoids, Adenoid Area, (Delawarr)	33.50	46.50	Freq: 20.0, 880.0
Mouth, Nose & Throat, Pharynx, Adenoids, Adenoid Area, (Delawarr)	76.00	36.00	Freq: 20.0, 880.0
Mouth, Nose & Throat, Pharynx, Constrictor Muscles, (Delawarr)	26.25	23.75	Freq: 20.0, 240.0
Mouth, Nose & Throat, Pharynx, Epiglottis, Thin Triangular Lidlike Piece of Cartilage, (Delawarr)	00.00	16.50	that Folds Back Over the Opening of the Windpipe During Swallowing to Prevent Foreign Matter to Enter the Lungs
Mouth, Nose & Throat, Pharynx, Epiglottis, Thin Triangular Lidlike Piece of Cartilage, (Delawarr)	15.50	43.50	that Folds Back Over the Opening of the Windpipe During Swallowing to Prevent Foreign Matter to Enter the Lungs
Mouth, Nose & Throat, Pharynx, Epiglottis, Thin Triangular Lidlike Piece of Cartilage, (Delawarr)	00.00	42.50	that Folds Back Over the Opening of the Windpipe During Swallowing to Prevent Foreign Matter to Enter the Lungs
Mouth, Nose & Throat, Pharynx, Eustachian Tubes, (Delawarr)	69.50	38.00	
Mouth, Nose & Throat, Pharynx, Mucous Coat, (Delawarr)	57.40	33.00	
Mouth, Nose & Throat, Pharynx, Nerves, (Delawarr)	33.00	40.75	
Mouth, Nose & Throat, Pharynx, Nerves, Glossopharyngeal Branch, (Delawarr)	00.00	26.50	
Mouth, Nose & Throat, Pharynx, Nerves, Glossopharyngeal Branch, (Delawarr)	75.00	16.00	
Mouth, Nose & Throat, Pharynx, Nerves, Of Sphenopalatine Ganglion, (Delawarr)	29.75	24.00	
Mouth, Nose & Throat, Pharynx, Stratified Epithelium, (Delawarr)	38.00	37.90	
Mouth, Nose & Throat, Pharynx, Tonsils, Lymphocytes, (Delawarr)	16.75	33.25	Freq: 20.0, 880.0
Mouth, Nose & Throat, Pharynx, Tonsils, Lymphoid Tissue, (Delawarr)	17.50	28.00	Freq: 20.0, 880.0
Mouth, Nose & Throat, Pharynx, Tonsils, Tonsillar Artery, (Delawarr)	25.40	15.75	Freq: 20.0, 880.0
Mouth, Nose & Throat, Pharynx, Tonsils, Tonsillar Plexus, (Delawarr)	26.25	24.75	Freq: 20.0, 880.0
Mouth, Nose & Throat, Pharynx, Uvula Palatina, (Delawarr)	29.75	24.25	
Mouth, Nose & Throat, Salivary Glands, (Delawarr)	30.75	36.25	
Mouth, Nose & Throat, Salivary Glands, Parotid Gland, Arteries, (Delawarr)	39.00	12.50	
Mouth, Nose & Throat, Salivary Glands, Parotid Gland, Jaws, (Delawarr)	39.00	24.00	

Patterns of Information	Rate Left	Rate Right	Energetic Processes
Mouth, Nose & Throat, Salivary Glands, Parotid Gland, Lymph Vessels, (Delawarr)	10.50	18.50	
Mouth, Nose & Throat, Salivary Glands, Parotid Gland, Nerves, (Delawarr)	56.60	52.00	
Mouth, Nose & Throat, Salivary Glands, Parotid Gland, Nerves, Auriculotemporal Branch, (Delawarr)	02.20	49.40	
Mouth, Nose & Throat, Salivary Glands, Parotid Gland, Nerves, Of Otic Ganglion, (Delawarr)	77.75	66.50	
Mouth, Nose & Throat, Salivary Glands, Parotid Gland, Veins, (Delawarr)	22.20	30.25	
Mouth, Nose & Throat, Salivary Glands, Sublingual Gland, (Delawarr)	16.00	23.50	
Mouth, Nose & Throat, Salivary Glands, Submaxillary Gland, (Delawarr)	22.75	14.50	
Muscular Structure, Aponeuroses, (Delawarr)	91.50	68.00	
Muscular Structure, Fasciae, (Delawarr)	73.50	81.50	Fascia Tissue Surrounds Organs & Muscles & Helps Keep Their Shape; Emotions Settle in these Tissues Which Tends To Lock Up the Fascia
Muscular Structure, Fasciae, Deep Fascia, (Delawarr)	64.50	57.10	Fascia Tissue Surrounds Organs & Muscles & Helps Keep Their Shape; Emotions Settle in these Tissues Which Tends To Lock Up the Fascia
Muscular Structure, Fasciae, Fatty Nodules, (Delawarr)	79.20	73.50	
Muscular Structure, Fasciae, Superficial Fascia, (Delawarr)	79.60	69.25	
Muscular Structure, Fasciae, Synovial Bursae, (Delawarr)	73.00	87.25	
Muscular Structure, Fasciae, Synovial Membrane, (Delawarr)	76.25	73.50	
Muscular Structure, Fasciae, Synovial Sheaths, (Delawarr)	68.50	61.30	
Muscular Structure, Ligaments, (Delawarr)	07.00	14.50	
Muscular Structure, Ligaments, (Delawarr)	58.75	65.00	
Muscular Structure, Ligaments, (Delawarr)	21.50	45.50	
Muscular Structure, Muscular Tissue, (Delawarr)	46.00	36.50	
Muscular Structure, Muscular Tissue, (Delawarr)	11.00	41.00	Tic, Twitch
Muscular Structure, Muscular Tissue, (Delawarr)	00.00	41.00	Balance 55.00-49.50 Indefinitely To Lower
Muscular Structure, Muscular Tissue, Blood Supply, (Delawarr)	58.25	43.50	
Muscular Structure, Muscular Tissue, Cardiac, (Delawarr)	86.50	48.40	
Muscular Structure, Muscular Tissue, Muscle Cells, (Delawarr)	58.50	51.00	Freq: 20.0, 240.0
Muscular Structure, Muscular Tissue, Muscle Fibers, (Delawarr)	56.75	61.50	Freq: 20.0, 240.0
Muscular Structure, Muscular Tissue, Muscle Fibers, Sarcolemma, (Delawarr)	57.25	22.25	Freq: 20.0, 240.0
Muscular Structure, Muscular Tissue, Muscle Sheath, (Delawarr)	53.50	53.25	Freq: 20.0, 240.0
Muscular Structure, Muscular Tissue, Muscle Sheath, (Delawarr)	00.00	42.50	Note E#, Sphincter; Freq: 324.0, 20.0, 240.0
Muscular Structure, Muscular Tissue, Nerve Supply, (Delawarr)	39.75	49.00	
Muscular Structure, Muscular Tissue, Nerve Supply, Motor Nerves, (Delawarr)	37.75	58.50	
Muscular Structure, Muscular Tissue, Nerve Supply, Motor Nerves, End Plates, (Delawarr)	26.50	28.00	With Depression Check Need To Increase or Decrease Acetylcholine
Muscular Structure, Muscular Tissue, Nerve Supply, Sensory Nerves, (Delawarr)	90.25	81.00	
Muscular Structure, Muscular Tissue, Nerve Supply, Sensory Nerves, Muscle Spindles, (Delawarr)	66.60	63.00	Freq: 20.0, 240.0
Muscular Structure, Muscular Tissue, Striped Or Voluntary, (Delawarr)	61.50	75.00	

Patterns of Information	Rate Left	Rate Right	Energetic Processes
Muscular Structure, Muscular Tissue, Striped Or Voluntary, Tendons, Ligaments, (Delawarr)	00.00	45.50	
Muscular Structure, Muscular Tissue, Unstriped Or Involuntary, (Delawarr)	00.00	57.00	
Muscular Structure, Muscular Tissue, Unstriped Or Involuntary, (Delawarr)	56.50	61.00	
Muscular Structure, Tendons, (Delawarr)	57.50	69.00	
Muscular Structure, Tendons, Nerve Supply, (Delawarr)	75.60	69.50	
Muscular Structure, Tendons, Sheaths, (Delawarr)	55.00	64.50	
Muscular, Strain, (Delawarr)	22.00	10.75	
Musculoskeletal & Connective Tissue Disorders, Osteoarthritis, (Delawarr)	23.25	23.25	Slow Progressive, Degenerative Joint Disease; Wearing Away of Cartilage in Joints; Freq: 43.3
Musculoskeletal & Connective Tissue Disorders, Ulcers, (Delawarr)	44.50	17.00	
Nerves, Impinged, (Delawarr)	21.80	23.50	
Neurologic Disorders, Muscular Atrophies & Dystrophies, (Delawarr)	16.50	15.50	A Wasting Away of Body Tissue; or Failure to Develop
Otorhinolaryngological Disorders, Nose & Paranasal Sinus Disorders, Acute Catarrh, (Delawarr)	39.50	33.50	Use Of 3.00-9.00 Adrenals is Effective. Freq. 43.3
Otorhinolaryngological Disorders, Nose & Paranasal Sinus Disorders, Chronic Catarrh, (Delawarr)	18.25	18.30	Use Of 3.00-9.00 Adrenals is Effective. Freq. 43.3
Pain, Increased at Night, Periostitis, (Delawarr)	10.75	14.50	
Pain, Increased by Movement, Rheumatism, Acute, (Delawarr)	37.75	44.00	
Pain, Relieved by Food & Pressure, Increased by Movement, Muscle Ruptured, (Delawarr)	28.75	23.50	
Pancreas, (Delawarr)	85.50	22.50	Chromium Needs to Be In Balance for Proper Function; Freq. 15.0
Pancreas, (Delawarr)	09.00	70.00	Chromium Needs to Be In Balance for Proper Function, Stimulates Enzyme Production; Freq. 15.0
Pancreas, Arteries, From Branch of Hepatic Artery, (Delawarr)	25.50	23.50	Freq. 15.0
Pancreas, Arteries, From Branch of Mesenteric Artery, (Delawarr)	29.75	23.25	Freq. 15.0
Pancreas, Arteries, From Splenic Artery, (Delawarr)	20.25	39.50	Freq. 15.0
Pancreas, Head of Pancreas, (Delawarr)	42.25	43.50	Freq. 15.0
Pancreas, Head of Pancreas, Anterior Surface, (Delawarr)	23.50	26.00	Freq. 15.0
Pancreas, Head of Pancreas, Posterior Surface, (Delawarr)	29.60	32.25	Freq. 15.0
Pancreas, Insulin-Free Part, (Delawarr)	43.00	53.00	Freq. 15.0
Pancreas, Lymph Vessels, (Delawarr)	24.25	37.50	Freq. 15.0
Pancreas, Nerves, From Splanchine Nerves, (Delawarr)	27.25	25.25	Freq. 15.0
Pancreas, Nerves, From Vagus Nerve, (Delawarr)	20.25	26.00	Freq. 15.0
Pancreas, Nerves, Through Splenic Plexus, (Delawarr)	29.60	32.50	Freq. 15.0
Pancreas, Organ, (Delawarr)	00.00	34.00	Note C#, Check Chromium Balance for Proper Function; Freq: 117.3, 15.0
Pancreas, Pancreatic Duct, (Delawarr)	27.50	21.75	Freq. 15.0
Pancreas, Pancreatic Duct, Accessory Pancreatic Duct, (Delawarr)	27.25	35.00	Freq. 15.0
Pancreas, Pancreatic Duct, Ampulla Of Bile Duct, (Delawarr)	27.75	31.75	Freq. 15.0
Pancreas, Structure, Alveoli, (Delawarr)	36.00	35.25	Freq. 15.0
Pancreas, Structure, Centro-acinar Cells, (Delawarr)	35.75	33.00	Freq. 15.0
Pancreas, Structure, Intercalcary Ducts, (Delawarr)	06.00	09.25	Freq. 15.0
Pancreas, Structure, Islets of Langerhans, (Delawarr)	08.00	91.00	Freq. 15.0
Pancreas, Structure, Islets of Langerhans, (Delawarr)	36.00	35.00	Freq. 15.0

Patterns of Information	Rate Left	Rate Right	Energetic Processes
Pancreas, Structure, Islets of Langerhans, Alpha Cells, Produces Glucagon,, (Delawarr)	32.00	24.00	Balances Blood Sugar; Freq. 15.0
Pancreas, Structure, Islets of Langerhans, Beta Cells, (Delawarr)	25.00	26.00	Freq. 15.0
Pancreas, Structure, Paranucleus, (Delawarr)	33.50	21.75	Freq. 15.0
Pancreas, Sugar-Free Part, Protein, (Delawarr)	06.00	04.00	Freq. 15.0
Pancreas, Tail of Pancreas, (Delawarr)	21.75	30.25	Freq. 15.0
Pancreas, Tail of Pancreas, Anterior Surface, (Delawarr)	27.50	30.00	Freq. 15.0
Pancreas, Tail of Pancreas, Inferior Surface, (Delawarr)	25.20	28.75	Freq. 15.0
Pancreas, Tail of Pancreas, Posterior Surface, (Delawarr)	32.80	22.50	Freq. 15.0
Pancreas, Tail of Pancreas, Superior Border, (Delawarr)	24.25	24.00	Freq. 15.0
Pancreas, Veins, (Delawarr)	22.00	31.75	Freq. 15.0
Paralysis, Agitans, (Delawarr)	11.00	19.00	Freq: 20.0
Perforation, (Delawarr)	56.00	53.00	
Pericardial Effusion, (Delawarr)	23.00	06.75	
Pericarditis, (Delawarr)	47.25	46.25	
Peritoneal Effusion, (Delawarr)	23.00	19.25	
Phlebitis, (Delawarr)	23.00	28.50	
Plant, Corn, (Delawarr)	16.50	16.00	
Psychiatric Disorders, Drug Addiction, Alcoholism, Alcoholic, Substance Abuse, (Delawarr)	28.75	26.00	
Psychological Shock, (Delawarr)	19.50	20.00	
Psychological Shock, Emotional, (Delawarr)	19.25	19.50	
Psychological Shock, Financial, (Delawarr)	19.75	20.00	
Psychological Shock, Mental Oppression, (Delawarr)	19.75	18.75	
Psychological Shock, Sexual, (Delawarr)	18.50	11.25	Power or Controlling Influence of One Person on Another That Can Not be Fought Off
Pulmonary Disorders, Airways Obstructions, Bronchial Asthma, (Delawarr)	40.00	18.50	Freq. 43.3
Pulmonary Disorders, Airways Obstructions, Bronchitis, (Delawarr)	22.00	17.00	
Pulmonary Disorders, Bronchiectasis, (Delawarr)	15.00	22.50	
Pulmonary Disorders, Pleural Disorders, Pleurisy, Often With Sharp Pain, Relieved, (Delawarr)	25.25	41.75	by Pressure, Increased by Movement Breathing or Coughing
Pulmonary Disorders, Pneumonia, (Delawarr)	32.25	11.00	Freq: 776.0
Radiation, Actinobacillosis, (Delawarr)	31.75	32.50	Infection by bacilli caused by Radiation
Respiratory System, Bronchi, (Delawarr)	04.00	74.00	
Respiratory System, Bronchi, (Delawarr)	92.00	82.00	
Respiratory System, Bronchi, (Delawarr)	00.00	83.00	
Respiratory System, Bronchi, Arteries, (Delawarr)	42.00	56.00	
Respiratory System, Bronchi, Bronchials, (Delawarr)	63.00	74.00	
Respiratory System, Bronchi, Cartilage Rings, (Delawarr)	47.25	65.20	
Respiratory System, Bronchi, Lymph Vessels, (Delawarr)	32.00	20.75	
Respiratory System, Bronchi, Lymph Vessels, Passing to Paratracheal Gland, (Delawarr)	29.00	18.25	
Respiratory System, Bronchi, Lymph Vessels, Passing to Pretracheal Gland, (Delawarr)	22.75	29.25	
Respiratory System, Bronchi, Mucous Membrane, (Delawarr)	46.00	46.00	
Respiratory System, Bronchi, Muscular Tissue, (Delawarr)	62.00	39.50	
Respiratory System, Bronchi, Nerves, (Delawarr)	29.75	28.50	
Respiratory System, Bronchi, Nerves, From Recurrent Laryngeal Nerves, (Delawarr)	39.75	31.25	

Patterns of Information	Rate Left	Rate Right	Energetic Processes
Respiratory System, Bronchi, Nerves, From Sympathetic Trunks, (Delawarr)	35.75	29.00	
Respiratory System, Bronchi, Nerves, From Vagus Nerve, (Delawarr)	37.00	07.75	
Respiratory System, Bronchi, Smooth Muscle, (Delawarr)	51.75	42.50	Freq: 20.0, 240.0
Respiratory System, Bronchi, Submucous Layer, (Delawarr)	40.75	24.50	
Respiratory System, Bronchi, Veins, (Delawarr)	50.40	42.25	
Respiratory System, Diaphragm, (Delawarr)	47.75	53.50	
Respiratory System, Diaphragm, (Delawarr)	33.00	58.00	
Respiratory System, Diaphragm, Diaphragm, Muscles, (Delawarr)	37.25	71.50	Freq: 20.0, 240.0
Respiratory System, Larynx, (Delawarr)	32.00	16.00	
Respiratory System, Larynx, (Delawarr)	61.50	47.75	
Respiratory System, Larynx, (Delawarr)	00.00	16.00	
Respiratory System, Larynx, Arteries, (Delawarr)	57.00	39.00	
Respiratory System, Larynx, Ligaments, (Delawarr)	32.25	39.00	
Respiratory System, Larynx, Membranes, (Delawarr)	42.00	28.00	
Respiratory System, Larynx, Mucous Membrane, (Delawarr)	66.00	49.50	
Respiratory System, Larynx, Muscles, (Delawarr)	38.50	36.20	Freq: 20.0, 240.0
Respiratory System, Larynx, Nerves, (Delawarr)	40.25	43.30	
Respiratory System, Larynx, The Nine Cartilages, (Delawarr)	59.50	33.25	
Respiratory System, Larynx, Veins, (Delawarr)	50.00	43.25	
Respiratory System, Lungs, (Delawarr)	88.25	08.60	Afected by Lyme Disease; Freq. 9.0
Respiratory System, Lungs, (Delawarr)	04.00	83.00	Affected by Lyme Disease; Freq. 9.0, 3672.0
Respiratory System, Lungs, Arteries, (Delawarr)	31.00	93.00	Freq. 9.0
Respiratory System, Lungs, Arteries, (Delawarr)	31.00	91.00	Freq. 9.0
Respiratory System, Lungs, Arteries, Pulmonary Artery, (Delawarr)	50.75	64.75	Freq. 9.0
Respiratory System, Lungs, Helpful to Pair with Large Intestine 8.00-72.00, (Delawarr)	92.00	83.00	Affected by Lyme Disease, Support for Neutralize Tobacco Toxins, Pair Color-White; Emotion-Sadness; Sound-Weeping; Freq. 9.0
Respiratory System, Lungs, Intercostal Muscles, External, (Delawarr)	52.50	39.50	Freq. 9.0, 20.0, 240.0
Respiratory System, Lungs, Intercostal Muscles, Internal, (Delawarr)	52.50	42.50	Freq. 9.0, 20.0, 240.0
Respiratory System, Lungs, Intrapulmonary Bronchi, (Delawarr)	85.75	63.50	Freq. 9.0
Respiratory System, Lungs, Intrapulmonary Bronchi, Alveoli, (Delawarr)	28.00	14.75	Freq. 9.0
Respiratory System, Lungs, Intrapulmonary Bronchi, Bronchial Muscles, (Delawarr)	50.00	35.00	Freq. 9.0, 20.0, 240.0
Respiratory System, Lungs, Intrapulmonary Bronchi, Mucous Membrane, (Delawarr)	34.50	34.25	Freq. 9.0
Respiratory System, Lungs, Intrapulmonary Bronchi, Outer Fibrous Coat, (Delawarr)	62.25	58.75	Freq. 9.0
Respiratory System, Lungs, Intrapulmonary Bronchi, Terminal Bronchioles, (Delawarr)	38.50	13.50	Freq. 9.0
Respiratory System, Lungs, Nerves, (Delawarr)	37.50	76.50	Freq. 9.0
Respiratory System, Lungs, Nerves, Broncho-constrictors From Vagus Nerve, (Delawarr)	39.25	46.00	Freq. 9.0
Respiratory System, Lungs, Nerves, From Sympathics, (Delawarr)	63.50	58.50	Freq. 9.0
Respiratory System, Lungs, Nerves, From Vagus Nerve, (Delawarr)	57.60	69.50	Freq. 9.0
Respiratory System, Lungs, Nerves, Pulmonary Plexus Anterior, (Delawarr)	49.00	39.50	Freq. 9.0
Respiratory System, Lungs, Nerves, Pulmonary Plexus Posterior, (Delawarr)	60.00	44.00	Freq. 9.0

Patterns of Information	Rate Left	Rate Right	Energetic Processes
Respiratory System, Lungs, Organ, (Delawarr)	00.00	47.00	Note A#, affected by lyme disease. Freq. 220.0, 9.0
Respiratory System, Lungs, Pulmonary Substance, (Delawarr)	39.50	30.75	Freq. 9.0
Respiratory System, Lungs, Pulmonary Substance, Lobules, (Delawarr)	59.00	42.50	Freq. 9.0
Respiratory System, Lungs, Serous Coat, (Delawarr)	74.74	28.50	Freq. 9.0
Respiratory System, Lungs, Sub-Serous Tissue, (Delawarr)	54.00	50.00	Freq. 9.0
Respiratory System, Lungs, Sub-Serous Tissue, (Delawarr)	54.00	40.00	Freq. 9.0
Respiratory System, Lungs, Thoracic Area, Thorax, Chest, Organ, (Delawarr)	00.00	53.00	Freq. 9.0
Respiratory System, Lungs, Veins, (Delawarr)	41.00	53.75	Freq. 9.0
Respiratory System, Mediastinum, (Delawarr)	40.50	62.50	
Respiratory System, Paranasal Sinuses, (Delawarr)	32.75	27.70	
Respiratory System, Paranasal Sinuses, Antrum, (Delawarr)	44.50	45.50	
Respiratory System, Paranasal Sinuses, Ethmoidal, (Delawarr)	33.90	43.00	
Respiratory System, Paranasal Sinuses, Frontal, (Delawarr)	36.75	37.00	
Respiratory System, Paranasal Sinuses, Frontal, (Delawarr)	00.00	50.00	
Respiratory System, Paranasal Sinuses, Frontal, (Delawarr)	00.00	03.00	
Respiratory System, Paranasal Sinuses, Maxillary, (Delawarr)	00.00	33.00	
Respiratory System, Paranasal Sinuses, Maxillary, (Delawarr)	32.50	20.00	
Respiratory System, Paranasal Sinuses, Nasal Mucous Membrane, (Delawarr)	43.00	46.00	
Respiratory System, Paranasal Sinuses, Sphenoidal, (Delawarr)	31.50	37.00	
Respiratory System, Pleurae, (Delawarr)	53.25	51.30	
Respiratory System, Pleurae, Arteries, (Delawarr)	26.75	43.00	
Respiratory System, Pleurae, Basement Membrane, (Delawarr)	53.25	72.00	
Respiratory System, Pleurae, Lymph Vessels, (Delawarr)	95.25	37.60	
Respiratory System, Pleurae, Nerves, (Delawarr)	79.00	69.25	
Respiratory System, Pleurae, Nerves, From Phrenic Nerve, (Delawarr)	42.75	33.50	
Respiratory System, Pleurae, Pleural Cavity, (Delawarr)	57.25	62.25	
Respiratory System, Pleurae, Serous Coat, (Delawarr)	46.25	44.50	
Respiratory System, Pleurae, Unstriped Muscle, (Delawarr)	34.50	34.50	Freq: 20.0, 240.0
Respiratory System, Trachea, (Delawarr)	31.75	75.40	
Respiratory System, Trachea, Cartilages, (Delawarr)	73.25	69.00	
Respiratory System, Trachea, Mucous Membrane, (Delawarr)	56.25	49.75	
Respiratory, Pleural Effusion, (Delawarr)	23.25	32.75	
Reticulo-Endothelial System, Cells Comprising Loose Reticulum In, Connective Tissues, (Delawarr)	28.00	30.20	
Reticulo-Endothelial System, Cells Comprising Loose Reticulum In, Endotheliocyte, (Delawarr)	42.50	50.25	Classmatocytes
Reticulo-Endothelial System, Cells Comprising Loose Reticulum In, Endothelium Lining, (Delawarr)	35.25	35.00	The Vascular Sinuses, Anterior Pituitary

Patterns of Information	Rate Left	Rate Right	Energetic Processes
Reticulo-Endothelial System, Cells Comprising Loose Reticulum In, Endothelium Lining, (Delawarr)	10.75	26.00	the Vascular sinuses
Reticulo-Endothelial System, Cells Comprising Loose Reticulum In, Endothelium Lining, (Delawarr)	29.00	30.00	The Vascular Sinuses, Liver, Kupffer's Cells
Reticulo-Endothelial System, Cells Comprising Loose Reticulum In, Endothelium Lining, (Delawarr)	17.00	20.00	The Vasuclar Sinuses, Lymph Glands
Reticulo-Endothelial System, Cells Comprising Loose Reticulum In, Endothelium Lining, (Delawarr)	14.00	20.75	The Vascular Sinuses, Spleen
Reticulo-Endothelial System, Cells Comprising Loose Reticulum In, Endothelium Lining, (Delawarr)	11.75	09.50	The Vascular Sinuses, In The Bone Marrow
Reticulo-Endothelial System, Cells Comprising Loose Reticulum In, Endothelium Lining, (Delawarr)	18.20	18.40	The Vascular Sinuses, Medulla Of Suprarenal Gland
Reticulo-Endothelial System, Cells Comprising Loose Reticulum In, Meninges, (Delawarr)	30.40	59.00	
Reticulo-Endothelial System, Cells Comprising Loose Reticulum In, Meningocytes, (Delawarr)	03.50	05.50	
Reticulo-Endothelial System, Cells Comprising Loose Reticulum In, Microglia Cells of, (Delawarr)	10.00	06.30	Neuroglia
Reticulo-Endothelial System, Cells Comprising Loose Reticulum In, the Blood, (Delawarr)	45.50	50.25	
Reticulo-Endothelial System, Cells Comprising Loose Reticulum In, the Blood, Large, (Delawarr)	36.75	18.50	Mononuclear Leucocytes
Reticulo-Endothelial System, Cells Of Bone Marrow, Red Marrow, (Delawarr)	25.10	18.50	
Reticulo-Endothelial System, Cells Of Bone Marrow, Red Marrow, Erythroblast, (Delawarr)	41.20	56.00	
Reticulo-Endothelial System, Cells Of Bone Marrow, Red Marrow, Erythrocyte, (Delawarr)	34.00	26.50	
Reticulo-Endothelial System, Cells Of Bone Marrow, Red Marrow, Myelocytes, (Delawarr)	41.75	39.50	
Reticulo-Endothelial System, Cells Of Bone Marrow, Red Marrow, Normoblast, (Delawarr)	29.50	52.25	
Reticulo-Endothelial System, Cells Of Bone Marrow, Red Marrow, Reticulocyte, (Delawarr)	57.00	64.00	
Reticulo-Endothelial System, Cells Of Bone Marrow, Reticulo-Endothelium, (Delawarr)	42.75	39.00	
Reticulo-Endothelial System, Cells Of Bone Marrow, Reticulo-Endothelium, Megakary-, (Delawarr)	93.60	85.75	ocytes, Found in Red Bone marrow then Evolves into thrombocytes or blood platelets at their death give off the chemical serotonin
Reticulo-Endothelial System, Cells Of Bone Marrow, Reticulo-Endothelium, Megakarycocytes, (Delawarr)	83.25	65.25	blasts
Reticulo-Endothelial System, Cells Of Bone Marrow, Reticulo-Endothelium, Plasma, (Delawarr)	71.30	59.00	Cells, Activated B Cell
Reticulo-Endothelial System, Cells Of Bone Marrow, Reticulo-Endothelium, Plasma-, (Delawarr)	72.00	68.25	blasts
Reticulo-Endothelial System, Cells Of Bone Marrow, Reticulo-Endothelium, Thrombocytes, Platelets,, (Delawarr)	83.50	95.50	As Fragments of megakarycoytes with life-span of 5-9 days, at death give off serotonin, With Depression Check Need To Increase Norepinephrins, Or Decrease Noradrenaline
Reticulo-Endothelial System, Cells Of Bone Marrow, Yellow Marrow, (Delawarr)	28.75	28.50	No longer produces blood cells
Reticulo-Endothelial System, Cells Of Bone Marrow, Yellow Marrow, Lymphoblasts, (Delawarr)	50.25	29.00	
Reticulo-Endothelial System, Cells Of Bone Marrow, Yellow Marrow, Lymphocytes, (Delawarr)	43.25	34.00	

Patterns of Information	Rate Left	Rate Right	Energetic Processes
Reticulo-Endothelial System, Cells Of Bone Marrow, Yellow Marrow, Metamyelocytes, (Delawarr)	33.20	26.50	
Reticulo-Endothelial System, Cells Of Bone Marrow, Yellow Marrow, Monoblasts, (Delawarr)	68.25	80.50	
Reticulo-Endothelial System, Cells Of Bone Marrow, Yellow Marrow, Monocytes, (Delawarr)	81.10	60.20	
Reticulo-Endothelial System, Cells Of Bone Marrow, Yellow Marrow, Myeloblasts, (Delawarr)	67.25	42.50	
Reticulo-Endothelial System, Cells Of Bone Marrow, Yellow Marrow, Myelocytes, (Delawarr)	65.25	48.25	
Reticulo-Endothelial System, Cells Of Bone Marrow, Yellow Marrow, Polymorphonuclears, (Delawarr)	31.25	27.60	Freq: 20.0, 880.0
Reticulo-Endothelial System, Cells Of Bone Marrow, Yellow Marrow, Promyelocytes, (Delawarr)	52.75	40.75	
Reticulo-Endothelial System, Motile Phagocytic Cells, (Delawarr)	13.25	15.00	
Reticulo-Endothelial System, Motile Phagocytic Cells, Blood Macrophages, (Delawarr)	21.20	09.50	
Rheumatism, Chronic, (Delawarr)	17.25	35.00	
Rheumatism, Gonorrheal, (Delawarr)	23.25	18.00	Freq: 600.0, 700.0
Sinuses, (Delawarr)	00.00	42.75	
Sinuses, Air Sinuses, (Delawarr)	05.50	33.00	Use 10.00-20.00 To Clear
Sinuses, Air Sinuses, (Delawarr)	26.00	33.00	
Sinuses, Air Sinuses, Antrum, (Delawarr)	36.40	31.25	
Sinuses, Air Sinuses, Ethmoid, (Delawarr)	28.40	47.00	
Sinuses, Air Sinuses, Frontal, (Delawarr)	42.75	39.00	
Sinuses, Air Sinuses, Mastoid, (Delawarr)	41.50	37.50	
Sinuses, Air Sinuses, Maxillary, (Delawarr)	00.00	23.25	
Sinuses, Air Sinuses, Maxillary, Lower Lungs, (Delawarr)	32.00	33.00	Freq. 9.0
Sinuses, Air Sinuses, Sphenoid, (Delawarr)	27.75	18.75	
Sinuses, Air Sinuses, Sphenoidal, Paranasal Sinuses, Respiratory System, (Delawarr)	00.00	37.50	
Sinuses, Venous Sinuses, (Delawarr)	11.25	17.50	
Sinuses, Venous Sinuses, Coronary, (Delawarr)	47.00	47.75	
Sinuses, Venous Sinuses, Dura Mater, Antero-Inferior Group, Cavernous, (Delawarr)	43.25	47.20	
Sinuses, Venous Sinuses, Dura Mater, Antero-Inferior Group, Basilar Network, (Delawarr)	46.50	47.20	
Sinuses, Venous Sinuses, Dura Mater, Antero-Inferior Group, Intercavernous, (Delawarr)	44.20	48.40	
Sinuses, Venous Sinuses, Dura Mater, Antero-Inferior Group, Middle Meningeal, (Delawarr)	50.60	58.00	
Sinuses, Venous Sinuses, Dura Mater, Antero-Inferior Group, Petrosal Inferior, (Delawarr)	49.50	49.00	
Sinuses, Venous Sinuses, Dura Mater, Antero-Inferior Group, Petrosal Superior, (Delawarr)	57.50	53.00	
Sinuses, Venous Sinuses, Dura Mater, Draining Blood From Brain, (Delawarr)	21.25	15.50	
Sinuses, Venous Sinuses, Dura Mater, Postero-Superior Group, Occipital, (Delawarr)	51.00	56.00	
Sinuses, Venous Sinuses, Dura Mater, Postero-Superior Group, Sagittal Inferior, (Delawarr)	58.40	50.20	
Sinuses, Venous Sinuses, Dura Mater, Postero-Superior Group, Sagittal Superior, (Delawarr)	54.25	58.00	

Patterns of Information	Rate Left	Rate Right	Energetic Processes
Sinuses, Venous Sinuses, Dura Mater, Postero-Superior Group, Sigmoid, (Delawarr)	45.90	40.25	
Sinuses, Venous Sinuses, Dura Mater, Postero-Superior Group, Straight, (Delawarr)	46.50	53.00	
Sinuses, Venous Sinuses, Dura Mater, Postero-Superior Group, Transverse, (Delawarr)	43.00	46.50	
Skeletal System, Bone Marrow, (Delawarr)	35.00	92.00	See Reticulo-Endothelial System
Skeletal System, Bone Marrow, (Delawarr)	01.00	04.40	See Reticulo-Endothelial System
Skeletal System, Bone, (Delawarr)	41.50	42.60	
Skeletal System, Bone, (Delawarr)	25.00	22.00	
Skeletal System, Bone, Blood Vessels, (Delawarr)	27.25	36.40	
Skeletal System, Bone, Blood Vessels, Canaliculi, (Delawarr)	90.75	82.40	
Skeletal System, Bone, Blood Vessels, Haversian Canals, (Delawarr)	38.25	40.00	
Skeletal System, Bone, Blood Vessels, Lacunae, (Delawarr)	78.50	80.25	
Skeletal System, Bone, Blood Vessels, Lamellae, (Delawarr)	47.75	50.10	
Skeletal System, Bone, Blood Vessels, Nutrient Artery, (Delawarr)	67.00	61.00	
Skeletal System, Bone, Bone Marrow, Organ, (Delawarr)	00.00	21.00	Note Ab#, See Reticulo-Endothelial System. Freq. 418.3
Skeletal System, Bone, Nerves, (Delawarr)	83.00	68.00	
Skeletal System, Bone, Osteogenetic Tissue, Young Bones Only, (Delawarr)	39.60	44.10	
Skeletal System, Bone, Periosteum, (Delawarr)	35.40	30.00	
Skeletal System, Cartilage, (Delawarr)	33.00	36.00	
Skeletal System, Cartilage, Cellular Cartilage, (Delawarr)	57.75	46.50	
Skeletal System, Cartilage, Elastic Fibrocartilage, (Delawarr)	48.50	51.25	
Skeletal System, Cartilage, Hyaline Cartilage, (Delawarr)	39.50	42.25	
Skeletal System, Cartilage, White Fibrocartilage, (Delawarr)	39.75	42.00	
Skeletal System, Joints, (Delawarr)	79.00	52.00	
Skeletal System, Joints, (Delawarr)	00.00	52.50	
Skeletal System, Vertebrae, (Delawarr)	34.75	34.25	See Spinal
Skeletal System, Vertebrae, Intervertebral Discs, (Delawarr)	38.75	40.25	
Skeletal, Dislocation, (Delawarr)	18.20	15.60	Use 3.00-80.00
Skeletal, Fracture, Often with Sharp Pain, (Delawarr)	12.00	12.00	
Skin, Hair & Nails, Alveoli, (Delawarr)	38.50	45.00	
Skin, Hair & Nails, Alveoli, Basement Membrane, (Delawarr)	39.25	49.50	
Skin, Hair & Nails, Alveoli, Epithelial Cells, (Delawarr)	40.00	44.50	
Skin, Hair & Nails, Alveoli, Larger Cells Containing Fat, (Delawarr)	42.00	36.50	
Skin, Hair & Nails, Alveoli, Sebum Cutaneum or Debris, (Delawarr)	34.40	41.00	
Skin, Hair & Nails, Corium, (Delawarr)	33.00	46.00	
Skin, Hair & Nails, Corium, Arteries, (Delawarr)	57.00	49.50	
Skin, Hair & Nails, Corium, Connective Tissue, (Delawarr)	40.50	41.00	
Skin, Hair & Nails, Corium, Connective Tissue, Papillary Layer, (Delawarr)	48.50	49.50	
Skin, Hair & Nails, Corium, Connective Tissue, Reticular Layer, (Delawarr)	43.00	41.00	
Skin, Hair & Nails, Corium, Ducts, (Delawarr)	36.50	42.50	

Patterns of Information	Rate Left	Rate Right	Energetic Processes
Skin, Hair & Nails, Corium, Lymph Vessels, (Delawarr)	50.75	52.20	
Skin, Hair & Nails, Corium, Muscular Fibers, (Delawarr)	58.40	46.50	
Skin, Hair & Nails, Corium, Nerves, (Delawarr)	34.00	28.25	
Skin, Hair & Nails, Corium, Sebaceous Glands, (Delawarr)	44.10	42.00	
Skin, Hair & Nails, Corium, Veins, (Delawarr)	45.50	53.00	
Skin, Hair & Nails, Cortex Of Hair, (Delawarr)	58.75	65.75	
Skin, Hair & Nails, Epidermis, (Delawarr)	70.00	02.50	
Skin, Hair & Nails, Epidermis, (Delawarr)	91.00	31.00	
Skin, Hair & Nails, Epidermis, (Delawarr)	00.00	42.00	
Skin, Hair & Nails, Epidermis, Germinative Zone, (Delawarr)	47.25	46.00	
Skin, Hair & Nails, Epidermis, Germinative Zone, Basal Cell Layer, (Delawarr)	56.00	49.50	
Skin, Hair & Nails, Epidermis, Germinative Zone, Prickle Cell Layer, (Delawarr)	42.00	37.40	
Skin, Hair & Nails, Epidermis, Horny Zone, (Delawarr)	59.60	52.50	
Skin, Hair & Nails, Epidermis, Horny Zone, Clear Layer, (Delawarr)	35.40	61.00	
Skin, Hair & Nails, Epidermis, Horny Zone, Granular Layer, (Delawarr)	50.50	33.00	
Skin, Hair & Nails, Epidermis, Horny Zone, Horny Layer, (Delawarr)	79.25	56.50	
Skin, Hair & Nails, Epithelial Cells, Epithelium, (Delawarr)	49.40	53.00	
Skin, Hair & Nails, Hair Follicle, (Delawarr)	00.00	85.00	
Skin, Hair & Nails, Hair Follicle, (Delawarr)	36.60	46.00	
Skin, Hair & Nails, Hair Follicle, Henle's Layer, (Delawarr)	47.00	40.25	
Skin, Hair & Nails, Hair Follicle, Hyaline Layer of Dermic Coat, (Delawarr)	64.50	63.20	
Skin, Hair & Nails, Hair Follicle, Outer Layer of Dermic Coat, (Delawarr)	33.60	46.00	
Skin, Hair & Nails, Hair, (Delawarr)	80.25	58.20	
Skin, Hair & Nails, Medulla Of Hair, (Delawarr)	78.50	66.00	
Skin, Hair & Nails, Nails, (Delawarr)	60.50	72.50	
Skin, Hair & Nails, Nails, Glomus, Important In Peripheral Circulation, (Delawarr)	62.50	64.75	
Skin, Hair & Nails, Nails, Lymphatics Of Nail Bed, (Delawarr)	52.75	48.25	
Skin, Hair & Nails, Nails, Matrix, (Delawarr)	46.20	55.50	
Skin, Hair & Nails, Nails, Nail Bed, (Delawarr)	65.50	55.10	
Skin, Hair & Nails, Nails, Nail Plate, (Delawarr)	49.50	55.25	
Skin, Hair & Nails, Nails, Nail Root, (Delawarr)	42.75	59.00	
Skin, Hair & Nails, Nails, Nail Wall, (Delawarr)	54.75	59.50	
Skin, Hair & Nails, Pigment, Melanin, (Delawarr)	58.20	53.20	
Skin, Hair & Nails, Sweat Glands, (Delawarr)	26.40	39.75	
Small Intestine, Arteries, Ileal, (Delawarr)	74.00	63.00	Freq. 10.0, 880.0
Small Intestine, Arteries, Jejunal, (Delawarr)	26.60	34.10	Freq. 10.0, 880.0
Small Intestine, Arteries, Mesentric Superior, (Delawarr)	26.40	45.90	Freq. 10.0, 880.0
Small Intestine, Helpful Paired with Heart 2.00-76.00, (Delawarr)	21.00	36.00	Pair Color-Red; Emotion-Happiness; Sound-Laughter; Freq. 10.0, 880.0
Small Intestine, Ileum, Lowest Part, Opening to Large Intestine, (Delawarr)	20.50	22.00	Freq. 10.0, 880.0
Small Intestine, Jejunum, (Delawarr)	12.25	23.00	Freq. 10.0, 880.0
Small Intestine, Lymph Vessels, (Delawarr)	22.50	35.50	Freq. 10.0, 880.0
Small Intestine, Mucous Membrane, (Delawarr)	51.75	42.25	Note C#; Freq: 281.0
Small Intestine, Mucous Membrane, Aggregated Lymphatic Nodules, Peyer's Patch, (Delawarr)	24.00	22.00	Found in the Ileum of Small Intestine, Important in releasing Antibodies & Absorption of Nutrients; Freq. 10.0, 880.0

Patterns of Information	Rate Left	Rate Right	Energetic Processes
Small Intestine, Mucous Membrane, Circular Folds, (Delawarr)	61.75	36.20	Freq. 10.0, 880.0
Small Intestine, Mucous Membrane, Solitary Lympathic Nodules, (Delawarr)	11.75	31.50	Freq. 10.0, 880.0
Small Intestine, Mucous Membrane, Villi, (Delawarr)	42.10	59.10	Freq. 10.0, 880.0
Small Intestine, Mucous Membrane, Villi, Artery Leaving Villus, (Delawarr)	81.50	17.60	Freq. 10.0, 880.0
Small Intestine, Mucous Membrane, Villi, Capillaries, (Delawarr)	17.25	11.75	Freq. 10.0, 880.0
Small Intestine, Mucous Membrane, Villi, Columnar Epithelium, (Delawarr)	22.75	25.25	Freq. 10.0, 880.0
Small Intestine, Mucous Membrane, Villi, Goblet Cells, (Delawarr)	26.25	27.75	Freq. 10.0, 880.0
Small Intestine, Mucous Membrane, Villi, Lacteals, (Delawarr)	03.80	16.50	Freq. 10.0, 880.0
Small Intestine, Mucous Membrane, Villi, Lymph Vessels, (Delawarr)	23.00	28.50	Freq. 10.0, 880.0
Small Intestine, Mucous Membrane, Villi, Muscle Cells, (Delawarr)	72.50	66.00	Freq. 10.0, 880.0, 20.0, 240.0
Small Intestine, Muscular Coat, (Delawarr)	13.75	26.50	Freq. 10.0, 880.0
Small Intestine, Muscular Coat, Circular Fibers, (Delawarr)	35.50	30.00	Freq. 10.0, 880.0
Small Intestine, Muscular Coat, Longitudinal Fibers, (Delawarr)	16.25	24.00	Freq. 10.0, 880.0
Small Intestine, Nerves, (Delawarr)	34.25	44.50	Freq. 10.0, 880.0
Small Intestine, Nerves, From Celiac Ganglion, (Delawarr)	60.75	70.50	Freq. 10.0, 880.0
Small Intestine, Nerves, From Meissner's Plexus From Which Fibers Also Pass to, (Delawarr)	49.50	49.25	Mucous Membrane; Freq. 10.0, 880.0
Small Intestine, Nerves, From Myenteric Plexus, (Delawarr)	46.10	56.60	Freq. 10.0, 880.0
Small Intestine, Nerves, From Plexus around Superior Mesenteric Artery, (Delawarr)	59.00	59.00	Freq. 10.0, 880.0
Small Intestine, Nerves, From Splanchnic Nerve, (Delawarr)	46.75	58.00	Freq. 10.0, 880.0
Small Intestine, Nerves, From Sympathetic, Inhibit Peristalsis, (Delawarr)	45.00	39.25	Freq. 10.0, 880.0
Small Intestine, Nerves, From Vagus Nerve, (Delawarr)	71.40	72.60	Freq. 10.0, 880.0
Small Intestine, Nerves, Parasympathetics, Augement Peristalsis & Secretion of, (Delawarr)	56.60	43.00	Gastric Juice; Freq. 10.0, 880.0
Small Intestine, Organ, (Delawarr)	00.00	82.00	Freq. 10.0, 880.0
Small Intestine, Plexus of Minute Branches, (Delawarr)	73.50	72.00	Freq. 10.0, 880.0
Small Intestine, Serous Coat, (Delawarr)	18.00	19.50	Freq. 10.0, 880.0
Small Intestine, Sub Mucous Coat, (Delawarr)	26.75	53.50	Freq. 10.0, 880.0
Small Intestine, Sub Mucous Coat, Blood Vessels, (Delawarr)	09.50	28.00	Freq. 10.0, 880.0
Small Intestine, Sub Mucous Coat, Lymphatics, (Delawarr)	32.50	37.10	Freq. 10.0, 880.0
Small Intestine, Sub Mucous Coat, Nerves, (Delawarr)	37.75	38.50	Freq. 10.0, 880.0
Small Intestine, Veins, (Delawarr)	35.50	44.50	Freq. 10.0, 880.0
Spasm, Involuntary Sudden Movement, (Delawarr)	33.25	03.50	
Spleen & Thymus, Ductless Glands, Thymus, (Delawarr)	59.00	77.00	Trains or Modifies T-Cells which regulate the humoral immune system; Freq: 20.0
Spleen & Thymus, Ductless Glands, Thymus, (Delawarr)	41.25	33.00	Freq: 20.0
Spleen & Thymus, Ductless Glands, Thymus, Organ, (Delawarr)	00.00	77.00	Trains or modifies T-Cells which regulate the humoral immune system; Freq: 20.0
Spleen & Thymus, Spleen, Arteries, Arterioles, (Delawarr)	39.75	45.25	Freq: 20.0

Patterns of Information	Rate Left	Rate Right	Energetic Processes
Spleen & Thymus, Spleen, Arteries, Branches In Spleen, (Delawarr)	37.00	55.00	Freq: 20.0
Spleen & Thymus, Spleen, Arteries, Splenic, (Delawarr)	36.00	66.50	Freq: 20.0
Spleen & Thymus, Spleen, External Serous Coat, (Delawarr)	13.25	17.50	Freq: 20.0
Spleen & Thymus, Spleen, Helpful Paired with stomach 77.00-84.00, (Delawarr)	04.00	95.00	Pair Color-Yellow; Emotion-Pensiveness; Sound-Singing; Freq: 20.0
Spleen & Thymus, Spleen, Internal Fibro-elastic Coat, (Delawarr)	22.25	26.50	Freq: 20.0
Spleen & Thymus, Spleen, Internal Fibro-elastic Coat, Splenic Cells, (Delawarr)	32.25	31.50	Freq: 20.0
Spleen & Thymus, Spleen, Internal Fibro-elastic Coat, Splenic Pulp, (Delawarr)	38.60	34.50	Freq: 20.0
Spleen & Thymus, Spleen, Internal Fibro-elastic Coat, Trabeculae, (Delawarr)	41.75	26.00	Freq: 20.0
Spleen & Thymus, Spleen, Left Side Under Arm, (Delawarr)	88.50	09.50	Freq: 20.0
Spleen & Thymus, Spleen, Left Side Under Arm, (Delawarr)	04.00	92.50	Note B#; Freq: 492.0
Spleen & Thymus, Spleen, Lymph Vessels, (Delawarr)	58.50	38.90	Freq: 20.0
Spleen & Thymus, Spleen, Lymph Vessels, Lymphatic Nodules, (Delawarr)	62.25	47.20	Freq: 20.0
Spleen & Thymus, Spleen, Nerves, (Delawarr)	44.00	35.00	Freq: 20.0
Spleen & Thymus, Spleen, Nerves, From Celiac Plexus, (Delawarr)	37.25	42.50	Freq: 20.0
Spleen & Thymus, Spleen, Organ, (Delawarr)	00.00	95.00	Freq. 20.0
Spleen & Thymus, Spleen, Peritoneum, (Delawarr)	91.00	31.00	Freq: 20.0
Spleen & Thymus, Spleen, Peritoneum, (Delawarr)	33.75	66.25	Freq: 20.0
Spleen & Thymus, Spleen, Veins, (Delawarr)	36.50	43.10	
Spleen & Thymus, Spleen, Veins, Tributaries, (Delawarr)	31.25	29.25	Freq: 20.0
Sprain, Trauma to a Joint that Causes Pain, (Delawarr)	28.00	48.50	
Stenosis, Constriction or Narrowing of a Passage, (Delawarr)	28.00	47.25	
Stomach, Arteries, (Delawarr)	25.00	31.00	Freq: 20.0
Stomach, Arteries, Gastric, Left, (Delawarr)	42.50	39.00	Freq: 20.0
Stomach, Arteries, Gastric, Right, (Delawarr)	29.90	34.00	Freq: 20.0
Stomach, Arteries, Gastroepiloic, Left, (Delawarr)	21.50	22.75	Freq: 20.0
Stomach, Arteries, Gastroepiloic, Right, (Delawarr)	31.00	42.60	Freq: 20.0
Stomach, Arteries, Plexus Of Larger Capillaries, (Delawarr)	28.00	33.10	Freq: 20.0
Stomach, Gastric Glands, (Delawarr)	23.75	23.25	Freq: 20.0
Stomach, Gastric Glands, Cardiac, (Delawarr)	37.75	36.20	Freq: 20.0
Stomach, Gastric Glands, Fundus, Source of Hydrochloric Acid (HC1), (Delawarr)	19.20	30.00	Freq: 20.0
Stomach, Gastric Glands, Pyloric, Source of Pepsin, (Delawarr)	29.00	23.80	Freq: 20.0
Stomach, Gastric Lymphatic Nodules, (Delawarr)	31.25	33.60	Freq: 20.0
Stomach, Helpful Paired with Spleen 4.00-95.00, (Delawarr)	77.00	84.00	Pair Color-Yellow; Emotion-Pensiveness; Sound-Singing; Freq: 20.0
Stomach, Inner Mucous Coat, (Delawarr)	24.25	22.75	Freq: 20.0
Stomach, Inner Mucous Coat, Columnar Epithelium, (Delawarr)	39.00	42.50	Freq: 20.0
Stomach, Inner Mucous Coat, Gland Ducts, (Delawarr)	45.25	49.00	Freq: 20.0
Stomach, Inner Mucous Coat, Goblet Cells, (Delawarr)	47.25	32.50	Freq: 20.0
Stomach, Inner Mucous Coat, Peptic Glands, (Delawarr)	45.00	39.75	Freq: 20.0

Patterns of Information	Rate Left	Rate Right	Energetic Processes
Stomach, Large Intestine, Omentum, (Delawarr)	00.00	52.50	Freq: 20.0
Stomach, Lymph Vessels, (Delawarr)	17.50	22.50	Freq: 20.0
Stomach, Muscular Coat, (Delawarr)	31.75	29.75	Freq: 20.0
Stomach, Muscular Coat, Circular Fibers, (Delawarr)	45.25	43.10	Freq: 20.0
Stomach, Muscular Coat, Longitudinal Fibers, (Delawarr)	39.60	31.50	Freq: 20.0
Stomach, Muscular Coat, Oblique Fibers, (Delawarr)	41.75	44.50	Freq: 20.0
Stomach, Nerves, (Delawarr)	27.75	21.00	Freq: 20.0
Stomach, Nerves, Parasympathetics, From Vagus Nerve, (Delawarr)	41.75	47.10	Freq: 20.0
Stomach, Nerves, Parasympathetics, Greater Anterior Gastric Nerve, (Delawarr)	42.00	48.20	Freq: 20.0
Stomach, Nerves, Parasympathetics, Greater Posterior Gastric Nerve, (Delawarr)	40.75	47.50	Freq: 20.0
Stomach, Nerves, Parasympathetics, Plexuses in Muscular Coat, (Delawarr)	40.00	44.50	Freq: 20.0
Stomach, Nerves, Parasympathetics, Plexuses in Sub Mucous Coat, (Delawarr)	46.00	37.50	Freq: 20.0
Stomach, Nerves, Sympathetic, from Left Phrenic Plexus, (Delawarr)	44.00	62.50	Freq: 20.0
Stomach, Nerves, Sympathetic, from Left Splanchnic Nerve, (Delawarr)	46.75	55.75	Freq: 20.0
Stomach, Nerves, Sympathetic, from Plexuses Around Gastric Artery, (Delawarr)	35.75	41.30	Freq: 20.0
Stomach, Nerves, Sympathetic, from Thoracic Sympathetic Trunk, (Delawarr)	48.60	53.50	Freq: 20.0
Stomach, Nerves, Sympathetics, from Celiac Plexus, (Delawarr)	33.90	51.60	Freq: 20.0
Stomach, Nerves, Sympathetics, from Lumbar Sympathetic Trunk, (Delawarr)	31.25	37.50	Freq: 20.0
Stomach, Nerves, Sympathetics, from Plexuses Around Epiloic Artery, (Delawarr)	34.00	58.20	Freq: 20.0
Stomach, Omentum, (Delawarr)	31.25	22.00	Freq: 20.0
Stomach, Organ, (Delawarr)	00.00	32.00	Note A#; Freq: 110.0, 20.0
Stomach, Pylorus, (Delawarr)	23.75	24.50	Freq: 20.0
Stomach, Pylorus, (Delawarr)	00.00	38.00	Freq: 20.0
Stomach, Serous Coat, (Delawarr)	47.50	28.25	Freq: 20.0
Stomach, Sub Mucous Coat, (Delawarr)	23.75	29.75	Freq: 20.0
Stomach, Sub Mucous Coat, Blood Vessels, (Delawarr)	43.20	34.50	Freq: 20.0
Stomach, Sub Mucous Coat, Lymphatics, (Delawarr)	19.75	36.50	Freq: 20.0
Stomach, Veins, (Delawarr)	26.00	32.00	Freq: 20.0
Stomach, Veins, Between Glands, (Delawarr)	39.20	46.25	Freq: 20.0
Stomach, Veins, Mesenteric, Superior, (Delawarr)	36.20	42.40	Freq: 20.0
Stomach, Veins, Portal, (Delawarr)	39.50	59.00	Freq: 20.0
Stomach, Veins, Splenic, (Delawarr)	35.50	38.75	Freq: 20.0
Subluxation, Partial Or Incomplete Separation, (Delawarr)	10.50	04.50	
Synovial Effusion, (Delawarr)	19.25	36.00	
Synovitus, Inflammation of the Synovial Membrane Surround the Joints, (Delawarr)	08.60	39.00	
Throat, Pharynx, Adenoids, Adenoid Area, (Delawarr)	22.00	26.25	Freq: 20.0, 880.0
Thrombosis, Blood Clot, (Delawarr)	36.50	25.50	
Toxins, Toxic State, (Delawarr)	32.20	21.00	
Trauma, (Delawarr)	10.25	24.00	Use with 68.00-49.00 to balance after surgery accident or a trauma to the body
Tumor, Benign, (Delawarr)	28.00	27.50	Freq: 666.0, 690.0, 740.0, 1840.0, 1998.0, 2000.0
Tumor, Benign, Cyst, (Delawarr)	25.00	28.00	Freq: 666.0, 690.0, 740.0, 1840.0, 1998.0, 2000.0
Tumor, Benign, Fibroma, (Delawarr)	27.25	11.00	Freq: 666.0, 690.0, 740.0, 1840.0, 1998.0, 2000.0

Patterns of Information	Rate Left	Rate Right	Energetic Processes
Tumor, Benign, Fibroma, Uterine, (Delawarr)	38.75	27.00	Freq: 666.0, 690.0, 740.0, 1840.0, 1998.0, 2000.0
Tumor, Benign, Polypus, (Delawarr)	49.25	14.00	Freq: 666.0, 690.0, 740.0, 1840.0, 1998.0, 2000.0
Tumor, Malignant, (Delawarr)	57.20	40.50	Freq: 666.0, 690.0, 740.0, 1840.0, 1998.0, 2000.0
Tumor, Malignant, Sarcoma, A Form of Cancer Which Develops from Connective Tissue, (Delawarr)	26.25	26.50	Freq: 2008.0
Tumor, Malignant, Sarcoma, A Form of Cancer Which Develops from Connective Tissue, (Delawarr)	26.25	26.50	Freq: 666.0, 690.0, 740.0, 1840.0, 1998.0, 2000.0
Ulcers, Varicose, (Delawarr)	21.75	37.50	
Urinary System, Bladder, (Delawarr)	26.25	20.50	
Urinary System, Bladder, (Delawarr)	08.00	67.00	
Urinary System, Bladder, (Delawarr)	48.00	75.00	
Urinary System, Bladder, (Delawarr)	13.00	67.00	
Urinary System, Bladder, Arteries, (Delawarr)	23.75	24.50	
Urinary System, Bladder, Female, (Delawarr)	19.00	14.00	
Urinary System, Bladder, Ligaments, (Delawarr)	22.00	19.40	
Urinary System, Bladder, Lymph Vessels, (Delawarr)	21.25	19.50	
Urinary System, Bladder, Male, (Delawarr)	21.00	15.00	
Urinary System, Bladder, Mucous Coat, (Delawarr)	21.50	24.25	
Urinary System, Bladder, Muscular Coat, (Delawarr)	41.75	25.00	
Urinary System, Bladder, Muscular Coat, External Longitudinal Layer, (Delawarr)	35.10	24.25	
Urinary System, Bladder, Muscular Coat, Internal Longitudinal Layer, (Delawarr)	26.75	23.00	
Urinary System, Bladder, Muscular Coat, Middle Circular Layer, (Delawarr)	30.00	24.50	
Urinary System, Bladder, Nerves, Parasympathetics Are Motor To Bladder Wall &, (Delawarr)	25.20	26.75	Inhibitory to The Sphincter
Urinary System, Bladder, Nerves, Sympathetics Are Motor To Sphincter & Inhibitory To, (Delawarr)	20.75	29.50	Muscles Of Bladder Wall
Urinary System, Bladder, Organ, (Delawarr)	00.00	13.00	Note F#, Balance 3.00-82.00; Freq. 352.0
Urinary System, Bladder, Serous Coat, (Delawarr)	29.00	26.00	
Urinary System, Bladder, Sub-mucous Coat, (Delawarr)	31.75	22.00	
Urinary System, Bladder, Universal, Helpful Paired with Kidney 3.00-82.00, (Delawarr)	39.00	30.00	Pair Color-Black; Emotion-Fear; Sound-Groaning; Freq. 8.0, 880.0
Urinary System, Bladder, Veins, (Delawarr)	66.75	29.60	
Urinary System, Body-Ureters, Left, (Delawarr)	45.00	28.25	
Urinary System, Body-Ureters, Right, (Delawarr)	40.75	12.50	
Urinary System, Calyces Minor, (Delawarr)	31.60	42.75	
Urinary System, Calyx Of Kidney, Cuplike Part or Cavity, (Delawarr)	52.25	50.50	Freq. 8.0, 880.0
Urinary System, Kidneys, (Delawarr)	32.00	23.00	Use for Arsenic, D D T, Earache, Edema, Flu, Meningococcus, Migraines, Nausea, Poisons, Polio, Scar Tissue, Sniffles; Freq. 8.0, 880.0
Urinary System, Kidneys, (Delawarr)	23.00	58.00	Use for Arsenic, D D T, Earache, Edema, Flu, Meningococcus, Migraines, Nausea, Poisons, Polio, Scar Tissue, Sniffles; Freq. 8.0, 880.0
Urinary System, Kidneys, Arteries, (Delawarr)	53.50	33.50	Freq. 8.0, 880.0
Urinary System, Kidneys, Cortical Substance, (Delawarr)	42.25	37.50	Freq. 8.0, 880.0
Urinary System, Kidneys, Fibrous Capsule, (Delawarr)	38.00	72.50	Freq. 8.0, 880.0
Urinary System, Kidneys, Helpful Paired with Urinary Bladder 39.00-30.00, (Delawarr)	03.00	82.00	Pair Color-Black; Emotion-Fear; Sound-Groaning; Freq. 8.0, 880.0
Urinary System, Kidneys, Lymph Vessels, (Delawarr)	35.50	19.75	Freq. 8.0, 880.0
Urinary System, Kidneys, Medullary Substance, (Delawarr)	62.00	66.50	Freq. 8.0, 880.0
Urinary System, Kidneys, Nerves, (Delawarr)	33.75	22.75	Freq. 8.0, 880.0

Patterns of Information	Rate Left	Rate Right	Energetic Processes
Urinary System, Kidneys, Organ, (Delawarr)	00.00	82.00	Balance for Arsenic, D D T, Earache, Edema, Flu, Meningococcus, Migraines, Nausea, Poisons, Polio, Scar Tissue, Sniffles; Freq. 8.0, 880.0
Urinary System, Kidneys, Organ, (Delawarr)	00.00	23.00	Note Eb#, E#, balance for arsenic, DDT, earache, edema, flu, meningococcus, migraines, nausea, poisons, polio, scar tissue, sniffles. Freq. 319.88, 330.0, 8.00, 880.0
Urinary System, Kidneys, Renal Capsules, Epithelial Lining, (Delawarr)	41.50	10.75	Freq. 8.0, 880.0
Urinary System, Kidneys, Renal Capsules, Glomerular Capsule, Bowman's Capsule, (Delawarr)	39.25	43.10	Freq. 8.0, 880.0
Urinary System, Kidneys, Renal Capsules, Glomerulus, (Delawarr)	46.00	65.25	Freq. 8.0, 880.0
Urinary System, Kidneys, Renal Capsules, Tubules, (Delawarr)	43.25	57.50	Freq. 8.0, 880.0
Urinary System, Kidneys, Uriniferous Tubules, (Delawarr)	53.75	19.75	Freq. 8.0, 880.0
Urinary System, Kidneys, Veins, (Delawarr)	38.00	12.50	Freq. 8.0, 880.0
Urinary System, Pelvis Of Ureter, (Delawarr)	34.25	33.25	
Urinary System, Urachus, (Delawarr)	34.50	34.25	
Urinary System, Ureters, (Delawarr)	43.75	48.00	
Urinary System, Ureters, Arteries, (Delawarr)	11.50	27.75	
Urinary System, Ureters, Fibrous Coat, (Delawarr)	47.00	23.50	
Urinary System, Ureters, Mucous Coat, (Delawarr)	27.25	25.50	
Urinary System, Ureters, Muscular Coat, (Delawarr)	35.25	25.25	
Urinary System, Ureters, Nerves, (Delawarr)	26.75	20.75	
Urinary System, Ureters, Organ, (Delawarr)	00.00	67.00	
Urinary System, Urethra, (Delawarr)	25.50	13.25	
Urinary System, Urethra, Mucous Membrane, (Delawarr)	28.25	31.60	
Urinary System, Urethra, Muscular Coat, (Delawarr)	29.00	28.25	
Urinary System, Urethra, Organ, (Delawarr)	00.00	86.00	
Urinary System, Urethra, Skene's Glands, (Delawarr)	40.75	29.25	
Urinary System, Urethra, Sphincter Urethrae, (Delawarr)	22.75	27.50	
Urinary System, Urethra, Spongy Erectile Tissue, (Delawarr)	20.25	19.75	
Valvular Incompetence, (Delawarr)	36.50	60.25	
Viral Diseases, Respiratory Viral Diseases, Coryza, (Delawarr)	22.00	02.00	Color Lemon, spread by Coughs & Sneezes, Scan & Use 96.00-0.00

Patterns of Information	Rate Left	Rate Right	Energetic Processes
Large Intestine, Mesentery, (Delawarr)	00.00	00.50	Freq. 10.0, 880.0
Central Nervous System, Fore Brain, Hypothalamus, Cowpers Glands, (Delawarr)	00.00	02.00	Should Equal GV, Check 22.00-0.00, 77.00-0.00, Balance Higher With Care
Respiratory System, Paranasal Sinuses, Frontal, (Delawarr)	00.00	03.00	
Central Nervous System, Hind Brain, Cerebellum, Fourth Ventricle, (Delawarr)	00.00	04.50	With Depression Check Need To Decrease Enkephalins
Central Nervous System, Brain Areas, Association Area, Frontal, (Delawarr)	00.00	05.00	Note Eb#, With Depression Check Need to Increase Enkephalins, Melatonin, Noradrenaline, Serotonin Or Decrease Dopamine. Freq. 315.8
Large Intestine, Ascending Colon, Organ, (Delawarr)	00.00	08.00	Note F#; Freq: 176.0, 20.0
Central Nervous System, Hind Brain, Pons, Mid, Medulla, (Delawarr)	00.00	10.00	Should Equal GV, With Depression Check Need To Decrease enkephalins
Ductless Glands, Thyroid Gland, Organ, (Delawarr)	00.00	11.00	Freq. 20.0, 160.0
Large Intestine, Sigmoid Flexure, (Delawarr)	00.00	12.00	Freq. 10.0, 880.0
Urinary System, Bladder, Organ, (Delawarr)	00.00	13.00	Note F#, Balance 3.00-82.00; Freq. 352.0
Ductless Glands, Thyroid Gland, (Delawarr)	00.00	14.50	Freq. 20.0, 160.0
Central Nervous System, Hind Brain, Pons Varolii, Fore, Medulla, (Delawarr)	00.00	15.00	With Depression Check Need To Decrease enkephalins
Respiratory System, Larynx, (Delawarr)	00.00	16.00	
Mouth, Nose & Throat, Pharynx, Epiglottis, Thin Triangular Lidlike Piece of Cartilage, (Delawarr)	00.00	16.50	that Folds Back Over the Opening of the Windpipe During Swallowing to Prevent Foreign Matter to Enter the Lungs
Lymphatic System, Lymph Glands, (Delawarr)	00.00	17.00	Freq. 10.0, 880.0
Eyes, Eyeball, (Delawarr)	00.00	18.00	Balance 9.50-10.50, often used for dreaming, Freq. 20.0
Eyes, Eyeball, Eyelids, Conjunctiva, (Delawarr)	00.00	19.00	Freq. 20.0
Skeletal System, Bone, Bone Marrow, Organ, (Delawarr)	00.00	21.00	Note Ab#, See Reticulo-Endothelial System. Freq. 418.3
Female Genital System, Ovaries, Organ, (Delawarr)	00.00	22.00	Freq. 1.0, 880.0
Male Genital System, Testes, Organ, (Delawarr)	00.00	22.00	Freq. 1.0, 880.0
Eyes, Eyeball, Fibrous Coat, Sclera, Retina, Optic Chiasma, (Delawarr)	00.00	22.50	Freq. 20.0
Urinary System, Kidneys, Organ, (Delawarr)	00.00	23.00	Note Eb#, E#, balance for arsenic, DDT, earache, edema, flu, meningococcus, migraines, nausea, poisons, polio, scar tissue, sniffles. Freq. 319.88, 330.0, 8.00, 880.0
Sinuses, Air Sinuses, Maxillary, (Delawarr)	00.00	23.25	
Gall Bladder & Biliary Ducts, Biliary Ducts, Bile Duct, (Delawarr)	00.00	23.50	
Mouth, Nose & Throat, Body-Nose, (Delawarr)	00.00	23.50	
Duodenum, (Delawarr)	00.00	25.00	Balance 16.00-25.00 if insect stings, migraines, parathion, poisons (arsenic), DDT, acidity in burns, or lindane is found in the duodenum
Male Genital System, Prostate Gland, Organ, (Delawarr)	00.00	26.00	Surrounds the neck of the bladder and urethra in males, it is part glandular and part muscular, gland secretes a thin opalescent, slightly alkaline fluid which is part of the semen. Freq. 20.0, 2000.0
Female Genital System, Uterus, Cervix Of Uterus, Neck of, (Delawarr)	00.00	26.00	
Breast, Mammary Glands, (Delawarr)	00.00	26.00	Amino Acid, Carnitine should be high in human breast milk
Mouth, Nose & Throat, Pharynx, Nerves, Glossopharyngeal Branch, (Delawarr)	00.00	26.50	
Central Nervous System, Ductless Glands, Pineal Body, (Delawarr)	00.00	27.00	With depression check need to increase melatonin, noradrenaline, serotonin or decrease dopamine, will be low for nervous conditions. Freq. 20.0, 5000.0
Eyes, Eyeball, Iris, (Delawarr)	00.00	27.50	Freq. 20.0
Eyes, Eyeball, Iris, (Delawarr)	00.00	28.00	Freq. 20.0
Liver, Organ, (Delawarr)	00.00	29.00	Note Eb#, Ab#. Freq. 317.83, 198.0
Stomach, Organ, (Delawarr)	00.00	32.00	Note A#; Freq: 110.0, 20.0
Respiratory System, Paranasal Sinuses, Maxillary, (Delawarr)	00.00	33.00	
Large Intestine, Pelvic Colon, (Delawarr)	00.00	33.75	Freq. 10.0, 880.0, 20.0
Pancreas, Organ, (Delawarr)	00.00	34.00	Note C#, Check Chromium Balance for Proper Function; Freq: 117.3, 15.0
Gall Bladder & Biliary Ducts, Gall Bladder, Organ, (Delawarr)	00.00	35.00	Note E#, Balance 13.00-64.00; Freq: 164.3, 330.0

Patterns of Information	Rate Left	Rate Right	Energetic Processes
Eyes, Eyeball, Vascular Coat, Choroid, (Delawarr)	00.00	36.50	Freq: 20.0
Sinuses, Air Sinuses, Sphenoidal, Paranasal Sinuses, Respiratory System, (Delawarr)	00.00	37.50	
Stomach, Pylorus, (Delawarr)	00.00	38.00	Freq: 20.0
Muscular Structure, Muscular Tissue, (Delawarr)	00.00	41.00	Balance 55.00-49.50 Indefinitely To Lower
Skin, Hair & Nails, Epidermis, (Delawarr)	00.00	42.00	
Muscular Structure, Muscular Tissue, Muscle Sheath, (Delawarr)	00.00	42.50	Note E#, Sphincter; Freq: 324.0, 20.0, 240.0
Large Intestine, Anus, (Delawarr)	00.00	42.50	Freq. 10.0, 880.0
Mouth, Nose & Throat, Pharynx, Epiglottis, Thin Triangular Lidlike Piece of Cartilage, (Delawarr)	00.00	42.50	that Folds Back Over the Opening of the Windpipe During Swallowing to Prevent Foreign Matter to Enter the Lungs
Sinuses, (Delawarr)	00.00	42.75	
Central Nervous System, Hind Brain, Cerebellum, (Delawarr)	00.00	43.25	With Depression Check Need To Decrease Enkephalins Area of Emotion
Mouth, Nose & Throat, Mouth, Tongue, Pharynx, Eustachian Tubes, Taste Buds, (Delawarr)	00.00	43.50	
Liver, Left Lobe, (Delawarr)	00.00	45.00	
Muscular Structure, Muscular Tissue, Striped Or Voluntary, Tendons, Ligaments, (Delawarr)	00.00	45.50	
Respiratory System, Lungs, Organ, (Delawarr)	00.00	47.00	Note A#, affected by lyme disease. Freq. 220.0, 9.0
Blood, Plasma, Volume, Organ, (Delawarr)	00.00	49.00	Note Eb#, balance 9.00-49.00 with caution. Freq. 321.9
Respiratory System, Paranasal Sinuses, Frontal, (Delawarr)	00.00	50.00	
Central Nervous System, Spinal Cord, Meninges, of Brain & Spinal Cord, Nerves, (Delawarr)	00.00	50.50	
Female Genital System, Vulva, (Delawarr)	00.00	51.50	
Central Nervous System, Fore Brain, Hypothalamus, Nerves, (Delawarr)	00.00	51.50	Balance 22.00-0.00, 77.00-0.00, Use Higher With Care
Skeletal System, Joints, (Delawarr)	00.00	52.50	
Stomach, Large Intestine, Omentum, (Delawarr)	00.00	52.50	Freq: 20.0
Respiratory System, Lungs, Thoracic Area, Thorax, Chest, Organ, (Delawarr)	00.00	53.00	Freq. 9.0
Female Genital System, Uterus, External Serous Coat, (Delawarr)	00.00	53.50	
Central Nervous System, Fore Brain, Thalamus, (Delawarr)	00.00	54.50	With Depression Check Need To Increase Norepinephrines In
Breast, Mammary Glands, (Delawarr)	00.00	56.00	
Muscular Structure, Muscular Tissue, Unstriped Or Involuntary, (Delawarr)	00.00	57.00	
Female Genital System, Uterus, Organ, (Delawarr)	00.00	58.00	
Female Genital System, Ovaries, (Delawarr)	00.00	59.50	Freq. 1.0, 880.0
Large Intestine, Cecum, Transverse Colon, (Delawarr)	00.00	60.00	Freq. 10.0, 880.0, 20.0
Large Intestine, Descending Colon, (Delawarr)	00.00	62.00	Freq. 10.0, 880.0, 20.0
Eyes, Eyeball, Lacrimal Apparatus, Lacrimal Ducts, (Delawarr)	00.00	64.00	Freq: 20.0
Large Intestine, Vermiform Appendix, Organ, (Delawarr)	00.00	66.00	Balance For Intestinal Gas (Flatulence), Typhoid Fever, Exanthematous Viral & Respiratory Viral Diseases, Malaria, Thyroid Imbalances, Eczema; Freq. 10.0, 880.0
Urinary System, Ureters, Organ, (Delawarr)	00.00	67.00	
Circulatory System, Veins, (Delawarr)	00.00	69.00	
Heart, Organ, (Delawarr)	00.00	70.00	Amino Acid Carnitine Highest providing fat oxidation which results in more energy; Freq: 20.0, 162.0
Circulatory System, Arteries, (Delawarr)	00.00	71.00	
Male Genital System, Penis, (Delawarr)	00.00	76.50	
Spleen & Thymus, Ductless Glands, Thymus, Organ, (Delawarr)	00.00	77.00	Trains or modifies T-Cells which regulate the humoral immune system; Freq: 20.0
Large Intestine, Hepatic Flexure, (Delawarr)	00.00	78.00	Freq. 10.0, 880.0
Central Nervous System, Fore Brain, (Delawarr)	00.00	78.50	With Depression Check Need to Increase Serotonin, Main Area of Brain
Female Genital System, Vagina, (Delawarr)	00.00	81.00	
Small Intestine, Organ, (Delawarr)	00.00	82.00	Freq. 10.0, 880.0

Patterns of Information	Rate Left	Rate Right	Energetic Processes
Urinary System, Kidneys, Organ, (Delawarr)	00.00	82.00	Balance for Arsenic, D D T, Earache, Edema, Flu, Meningococcus, Migraines, Nausea, Poisons, Polio, Scar Tissue, Sniffles; Freq. 8.0, 880.0
Central Nervous System, Brain Areas, Association Area, Frontal, (Delawarr)	00.00	82.75	
Respiratory System, Bronchi, (Delawarr)	00.00	83.00	
Skin, Hair & Nails, Hair Follicle, (Delawarr)	00.00	85.00	
Male Genital System, Spermatic Cord, (Delawarr)	00.00	86.00	High concentrations of amino acid carnitine, which provides energy for sperms motility
Urinary System, Urethra, Organ, (Delawarr)	00.00	86.00	
Female Genital System, Uterus, Cervix Of Uterus, Neck of, (Delawarr)	00.00	86.00	
Ductless Glands, Pituitary Body, (Delawarr)	00.00	87.00	With Depression Check Need to Increase Enkephalins, Serotonin
Large Intestine, Descending Colon, (Delawarr)	00.00	90.00	Freq. 10.0, 880.0, 20.0
Central Nervous System, Fore Brain, Hypothalamus, (Delawarr)	00.00	90.50	
Ductless Glands, Parathyroid Gland, (Delawarr)	00.00	92.00	
Spleen & Thymus, Spleen, Organ, (Delawarr)	00.00	95.00	Freq. 20.0
Circulatory System, Veins, (Delawarr)	00.00	97.00	
Ductless Glands, Thyroid Gland, (Delawarr)	00.00	98.00	Freq. 20.0, 160.0
Mouth, Nose & Throat, Mouth, Tongue, Circumvallate Papillae, (Delawarr)	00.40	24.50	
Skeletal System, Bone Marrow, (Delawarr)	01.00	04.40	See Reticulo-Endothelial System
Central Nervous System, Cranial Nerves, Spinal Or Peripheral Nerves, Cervical 8 Pairs, (Delawarr)	01.00	21.00	
Female Genital System, Ovaries, Tubal Extremity, (Delawarr)	01.50	56.25	Freq. 1.0, 880.0
Central Nervous System, Hind Brain, Cerebellum, Left, (Delawarr)	02.00	29.00	With Depression Check Need To Decrease Enkephalins Area Of Emotions
Central Nervous System, Hind Brain, Nerves, Cerebellum, Right, (Delawarr)	02.00	31.00	With Depression Check Need To Decrease Enkephalins Area of Emotion
Heart, Helpful Paired with Small Intestine 21.00-36.00, (Delawarr)	02.00	76.00	Pair Color-Red; Emotion-Happiness; Sound-Laughter; Freq: 20.0, 162.0
Male Genital System, Scrotum, (Delawarr)	02.00	98.00	
Mouth, Nose & Throat, Salivary Glands, Parotid Gland, Nerves, Auriculotemporal Branch, (Delawarr)	02.20	49.40	
Large Intestine, Vermiform Appendix, (Delawarr)	03.00	08.00	Balance For Intestinal Gas (Flatulence), Typhoid Fever, Exanthematous Viral & Respiratory Viral Diseases, Malaria, Thyroid Imbalances, Eczema; Freq. 10.0, 880.0
Ductless Glands, Suprarenal (Adrenal) Glands, (Delawarr)	03.00	09.00	Balance For Anemia, Catarrh, Edema, Exhaustion, Mononucleosis, Multiple Sclerosis, Poisons, Shock, Tuberculosis, Toxicity
Mouth, Nose & Throat, Mouth, Gums, (Delawarr)	03.00	25.00	
Female Genital System, Ovaries, (Delawarr)	03.00	54.00	Freq. 1.0, 880.0
Urinary System, Kidneys, Helpful Paired with Urinary Bladder 39.00-30.00, (Delawarr)	03.00	82.00	Pair Color-Black; Emotion-Fear; Sound-Groaning; Freq. 8.0, 880.0
Lymphatic System, Lymph Glands, (Delawarr)	03.00	100.00	Freq. 10.0, 880.0
Central Nervous System, Spinal Cord, Meninges, Arachnoid Villi, (Delawarr)	03.25	46.25	
Reticulo-Endothelial System, Cells Comprising Loose Reticulum In, Meningocytes, (Delawarr)	03.50	05.50	
Heart, Pericardium, Serous Sac, (Delawarr)	03.50	21.75	Freq: 20.0, 162.0
Breast, Mammary Glands, Nipple, Areolar Glands, (Delawarr)	03.75	09.50	
Small Intestine, Mucous Membrane, Villi, Lacteals, (Delawarr)	03.80	16.50	Freq. 10.0, 880.0
Internal Ear, Wax or Cerumen, (Delawarr)	04.00	06.00	Use 40.00-6.00; Freq: 20.0, 880.0
Acid, Uric, Crystals (Delawarr)	04.00	08.00	Deposit of this Mineral Could Cause Arthritic Problems, Balance 5.00-91.00, Check Gout 19.50-33.75 or 85.00-56.00
Respiratory System, Bronchi, (Delawarr)	04.00	74.00	
Respiratory System, Lungs, (Delawarr)	04.00	83.00	Affected by Lyme Disease; Freq. 9.0, 3672.0
Spleen & Thymus, Spleen, Left Side Under Arm, (Delawarr)	04.00	92.50	Note B#; Freq: 492.0

Patterns of Information	Rate Left	Right	Energetic Processes
Spleen & Thymus, Spleen, Helpful Paired with stomach 77.00-84.00, (Delawarr)	04.00	95.00	Pair Color-Yellow; Emotion-Pensiveness; Sound-Singing; Freq: 20.0
Mouth, Nose & Throat, Mouth, Teeth, (Delawarr)	04.00	99.00	
Heart, Pericardium, (Delawarr)	04.50	12.50	Freq: 20.0, 162.0
Heart, Veins, Posterior Vein Of Left Ventricle, (Delawarr)	04.75	08.50	Freq: 20.0, 162.0
Central Nervous System, Autonomic Nervous System, Sympathetic System, (Delawarr)	05.00	60.00	
Central Nervous System, Autonomic Nervous System, Sympathetic System, (Delawarr)	05.00	61.50	
Central Nervous System, (Delawarr)	05.00	62.00	
Male Genital System, Prostate Gland, (Delawarr)	05.00	91.00	Surrounds the Neck of the Bladder & Urethra In Males, It is Part Glandular & part Muscular, Gland Secretes a Thin Opalescent, Slightly Alkaline Fluid Which Is Part of Semen; Freq: 20.0, 2000.0
Sinuses, Air Sinuses, (Delawarr)	05.50	33.00	Use 10.00-20.00 To Clear
Pancreas, Sugar-Free Part, Protein, (Delawarr)	06.00	04.00	Freq. 15.0
Pancreas, Structure, Intercalcary Ducts, (Delawarr)	06.00	09.25	Freq. 15.0
Ductless Glands, Thyroid Gland, (Delawarr)	06.00	81.00	Use to Eliminate 82.00-XX. Freq. 20.0, 160.0
Ductless Glands, Pituitary Body, (Delawarr)	06.00	87.00	
Male Genital System, Penis, (Delawarr)	06.00	88.00	
Heart, Ventricles, Right, (Delawarr)	06.60	11.30	Freq: 20.0, 162.0
Muscular Structure, Ligaments, (Delawarr)	07.00	14.50	
Hepatic & Biliary Disorders, Calculus, Biliary, (Delawarr)	07.00	35.00	Abnormal Stony Mass In Bile Tract
Liver, Hepatic Artery, Conveying Arterial Blood To, Interlobular Plexuses, (Delawarr)	07.25	20.50	
Liver, Filter Factor, (Delawarr)	07.50	29.00	To Heal, Regenerate, Make Better
Heart, Veins, Middle Cardiac, (Delawarr)	07.75	06.50	Freq: 20.0, 162.0
External & Physical Agent Disorders, Insulin Shock, (Delawarr)	07.75	12.50	
Female Genital System, Uterus, Ligaments, Anterior Ligament, (Delawarr)	08.00	05.50	
Heart, Valves, Mitral, (Delawarr)	08.00	10.40	Freq: 20.0, 162.0
Heart, Valves, Aortic, (Delawarr)	08.00	10.60	Freq: 20.0, 162.0
Female Genital System, Uterus, (Delawarr)	08.00	23.00	
Urinary System, Bladder, (Delawarr)	08.00	67.00	
Large Intestine or Colon, Helpful to Pair with Lungs 92.00-83.00, (Delawarr)	08.00	72.00	Pair Color-White; Emotion-Sadness; Sound-Weeping; Freq. 9.0
Fever, Hay, Asthma, (Delawarr)	08.00	80.00	Freq. 43.3, 20.0, 880.0
Pancreas, Structure, Islets of Langerhans, (Delawarr)	08.00	91.00	Freq. 15.0
Heart, Veins, Coronary Sinus, (Delawarr)	08.25	10.00	Freq: 20.0, 162.0
Metabolism, Imbalance, (Delawarr)	08.25	29.25	
Large Intestine, Rectum, Organ, (Delawarr)	08.50	90.00	Use 55.00-96.00, Then 40.00-96.00; Freq. 10.0, 880.0
Synovitus, Inflammation of the Synovial Membrane Surround the Joints, (Delawarr)	08.60	39.00	
Breast, Mammary Glands, Gland Tissue, Colostrum Corpuscles, (Delawarr)	08.75	09.00	
Diseases, Septicemia, Bacteria in the Blood, (Delawarr)	08.75	35.00	Colloidal Silver is Effective
Heart, Valves, Pulmonary, (Delawarr)	09.00	05.60	Freq: 20.0, 162.0
Ductless Glands, Parathyroid Gland, Connective Tissue, Oxyphil Granules, (Delawarr)	09.00	30.00	
Blood, Plasma, General Vitality, (Delawarr)	09.00	49.00	Use With Caution, With Depression Check Need To Decrease Enkephalins, Melatonin, Serotonin
Pancreas, (Delawarr)	09.00	70.00	Chromium Needs to Be In Balance for Proper Function, Stimulates Enzyme Production; Freq. 15.0
External & Physical Agent Disorders, Physical Shock, (Delawarr)	09.25	13.00	
Small Intestine, Sub Mucous Coat, Blood Vessels, (Delawarr)	09.50	28.00	Freq. 10.0, 880.0
Mouth, Nose & Throat, Nose, Upper Nasal Cartilage, (Delawarr)	09.75	16.50	Freq: 20.0, 880.0

Patterns of Information	Rate Left	Rate Right	Energetic Processes
Arthritis, Degenerative (Delawarr)	09.75	19.00	Check For Mineral Deposits, Calcium, Carbon-Stop Massive Carbohydrate Intake, Silicon, Sodium-Use Sun-Dried Sea Salt, Uric Acid; Freq: 43.3
Breast, Mammary Glands, Gland Tissue, Blood Vessels, (Delawarr)	09.75	22.40	
Reticulo-Endothelial System, Cells Comprising Loose Reticulum In, Microglia Cells of, (Delawarr)	10.00	06.30	Neuroglia
Female Genital System, Uterus, (Delawarr)	10.00	61.00	
Heart, Myocardium, Fibrous Muscular Rings, (Delawarr)	10.20	22.75	Freq: 20.0, 162.0
Trauma, (Delawarr)	10.25	24.00	Use with 68.00-49.00 to balance after surgery accident or a trauma to the body
Breast, Mammary Glands, Gland Tissue, Areolar, (Delawarr)	10.40	07.25	
Heart, Valves, Chordae Tendineae, (Delawarr)	10.40	12.25	Freq: 20.0, 162.0
Subluxation, Partial Or Incomplete Separation, (Delawarr)	10.50	04.50	
Mouth, Nose & Throat, Mouth, Tongue, Fungiform Papillae, (Delawarr)	10.50	07.80	
Heart, Veins, Right Marginal, (Delawarr)	10.50	08.25	Freq: 20.0, 162.0
Mouth, Nose & Throat, Salivary Glands, Parotid Gland, Lymph Vessels, (Delawarr)	10.50	18.50	
Large Intestine, Ileocecal Valve, Frenula, (Delawarr)	10.50	39.50	Freq. 10.0, 880.0
Heart, Myocardium, Fibrous Muscular Rings, Longitudinal, (Delawarr)	10.75	08.50	Freq: 20.0, 162.0
Pain, Increased at Night, Periostitis, (Delawarr)	10.75	14.50	
External & Physical Agent Disorders, Shock Therapy, Thermal, (Delawarr)	10.75	14.50	
Heart, Ventricles, (Delawarr)	10.75	16.25	Freq: 20.0, 162.0
Warts, Circumscribed Cutaneous Elevation resulting from Hypertrophy of the Papillae &, (Delawarr)	10.75	20.50	Epidermis, Caused By The Papillomavirus, Balance Acid/Alkalinity & NA/CL
Reticulo-Endothelial System, Cells Comprising Loose Reticulum In, Endothelium Lining, (Delawarr)	10.75	26.00	the Vascular sinuses
Heart, Veins, (Delawarr)	11.00	08.50	Freq: 20.0, 162.0
Paralysis, Agitans, (Delawarr)	11.00	19.00	Freq: 20.0
Muscular Structure, Muscular Tissue, (Delawarr)	11.00	41.00	Tic, Twitch
Mouth, Nose & Throat, Body-Mouth, (Delawarr)	11.00	81.00	
Female Genital System, Vagina, Mucous Membrane, (Delawarr)	11.10	10.00	
Duodenum, Mucous Membrane, Villi, Columnar Epithelium, (Delawarr)	11.10	12.50	
Large Intestine, Structure Of, Mucous Membrane, (Delawarr)	11.25	14.75	Freq. 10.0, 880.0
Sinuses, Venous Sinuses, (Delawarr)	11.25	17.50	
Heart, Atria Septum, (Delawarr)	11.25	20.50	Freq: 20.0, 162.0
Heart, Nerves, (Delawarr)	11.25	20.75	Freq: 20.0, 162.0
Degeneration, (Delawarr)	11.25	41.50	
Heart, Valves, Tricuspid, (Delawarr)	11.50	10.00	Freq: 20.0, 162.0
Heart, Endocardium, (Delawarr)	11.50	11.10	Freq: 20.0, 162.0
Mouth, Nose & Throat, Nose, Respiratory Region, Nerves, (Delawarr)	11.50	12.50	Freq: 20.0, 880.0
Urinary System, Ureters, Arteries, (Delawarr)	11.50	27.75	
Heart, Arteries, Right Coronary, (Delawarr)	11.60	10.10	Freq: 20.0, 162.0
Female Genital System, Ovaries, Suspensory Ligament Of, (Delawarr)	11.60	35.30	Freq. 1.0, 880.0
Reticulo-Endothelial System, Cells Comprising Loose Reticulum In, Endothelium Lining, (Delawarr)	11.75	09.50	The Vascular Sinuses, In The Bone Marrow
Heart, Auricles, Auriculo Ventricular Ring, (Delawarr)	11.75	21.00	Freq: 20.0, 162.0

Patterns of Information	Rate Left	Rate Right	Energetic Processes
Small Intestine, Mucous Membrane, Solitary Lympathic Nodules, (Delawarr)	11.75	31.50	Freq. 10.0, 880.0
Mouth, Nose & Throat, Mouth, Tongue, Extrinsic Muscles, (Delawarr)	11.75	46.25	Freq: 20.0, 240.0
Skeletal, Fracture, Often with Sharp Pain, (Delawarr)	12.00	12.00	
Heart, Sino-Auricular Node, (Delawarr)	12.00	19.50	Freq: 20.0, 162.0
Heart, Nerves, Cardiac Plexus, (Delawarr)	12.00 00.00	22.50 06.00	Freq: 20.0, 162.0
Liver, Liver Factor, (Delawarr)	12.00	29.00	To Build, Increase, Enhance
Large Intestine, Rectum, (Delawarr)	12.00	35.00	Freq. 10.0, 880.0
Heart, Veins, Great Cardiac, (Delawarr)	12.20	09.30	Freq: 20.0, 162.0
Small Intestine, Jejunum, (Delawarr)	12.25	23.00	Freq. 10.0, 880.0
Heart, Pericardium, Nerves, (Delawarr)	12.40	28.20	Freq: 20.0, 162.0
Heart, Veins, Coronary Sinus, Valve Of Coronary Sinus, (Delawarr)	12.50	08.30	Freq: 20.0, 162.0
Acid, Uric, Crystals (Delawarr)	12.50	20.50	
External & Physical Agent Disorders, Poisoning Shock, (Delawarr)	12.50	34.25	
Breast, Mammary Glands, (Delawarr)	12.50	56.00	
Mouth, Nose & Throat, Mouth, Tongue, Mucous Membrane, (Delawarr)	12.50	59.00	
Large Intestine, Transverse Colon, Phrenicocolic Ligament, (Delawarr)	13.00	06.00	Freq. 10.0, 880.0, 20.0
Heart, Veins, Oblique Vein Of Left Atrium, (Delawarr)	13.00	21.00	Freq: 20.0, 162.0
Male Genital System, Prostate Gland, (Delawarr)	13.00	26.00	Surrounds the Neck of the Bladder & Urethra In Males, It is Part Glandular & part Muscular, Gland Secretes a Thin Opalescent, Slightly Alkaline Fluid Which Is Part of Semen; Freq: 20.0, 2000.0
Gall Bladder & Biliary Ducts, Gall Bladder, Helpful Paired with Liver, (Delawarr)	13.00	64.00	Use for Constipation, Lindane, Various Colds, Intestinal Worms, Pair Color-Green; Emotion-Anger; Sound-Shouting
Urinary System, Bladder, (Delawarr)	13.00	67.00	
Heart, Myocardium, Fibrous Muscular Rings, Transverse, (Delawarr)	13.10	04.00	Freq: 20.0, 162.0
Heart, Pericardium, Arteries, (Delawarr)	13.25	12.75	Freq: 20.0, 162.0
External & Physical Agent Disorders, Radioactive Fallout, Atomic, (Delawarr)	13.25	13.50	Potassium Chloride, Kelp Helpful
Reticulo-Endothelial System, Motile Phagocytic Cells, (Delawarr)	13.25	15.00	
Spleen & Thymus, Spleen, External Serous Coat, (Delawarr)	13.25	17.50	Freq: 20.0
Large Intestine, Vermiform Appendix, Circular Muscular Fibers, (Delawarr)	13.50	07.20	Freq. 10.0, 880.0
Heart, Endocardium, Connective Tissue, (Delawarr)	13.50	14.90	Freq: 20.0, 162.0
Heart, Ventricles, Left, (Delawarr)	13.50	16.50	Freq: 20.0, 162.0
Mouth, Nose & Throat, Esophagus, Muscule Fibers, Longitudinal, (Delawarr)	13.50	19.00	
Male Genital System, Prostate Gland, (Delawarr)	13.50	26.00	Surrounds the Neck of the Bladder & Urethra In Males, It is Part Glandular & part Muscular, Gland Secretes a Thin Opalescent, Slightly Alkaline Fluid Which Is Part of Semen; Freq: 20.0, 2000.0
Female Genital System, Uterus, (Delawarr)	13.50	34.00	
Heart, Veins, Left Marginal, (Delawarr)	13.60	15.30	Freq: 20.0, 162.0
Heart, Arteries, Right Coronary, Marginal Branch, (Delawarr)	13.75	10.50	Freq: 20.0, 162.0
Heart, Bundle of His, (Delawarr)	13.75	12.00	Freq: 20.0, 162.0
Male Genital System, Prostate Gland, Veins, (Delawarr)	13.75	21.50	Freq: 20.0, 2000.0
Small Intestine, Muscular Coat, (Delawarr)	13.75	26.50	Freq. 10.0, 880.0
Breast, Mammary Glands, Gland Tissue, Lactiferous Sinuses, (Delawarr)	14.00	11.25	
Eyes, Eyeball, Orbital Cavity, Rectus Medial, (Delawarr)	14.00	20.00	Freq: 20.0

Patterns of Information	Rate Left	Rate Right	Energetic Processes
Reticulo-Endothelial System, Cells Comprising Loose Reticulum In, Endothelium Lining, (Delawarr)	14.00	20.75	The Vascular Sinuses, Spleen
Blood, Blood Corpuscles (Delawarr)	14.00	36.20	
Gall Bladder & Biliary Ducts, Gall Bladder, (Delawarr)	14.00	64.00	Use For Constipation, Lindane, Various Colds, Intestinal Worms, Pair Color-Green; Emotion-Anger; Sound-Shouting
Heart, Lymph Vessels, Superficial Plexus, (Delawarr)	14.20	14.00	Freq: 20.0, 162.0
Heart, Auricles, (Delawarr)	14.20	14.60	Freq: 20.0, 162.0
Heart, Veins, Anterior Cardiac, (Delawarr)	14.20	16.50	Freq: 20.0, 162.0
Heart, Nerves, Cardiac Plexus, Superficial, (Delawarr)	14.50	15.00	Freq: 20.0, 162.0
Heart, Atria, (Delawarr)	14.75	16.75	Freq: 20.0, 162.0
Mouth, Nose & Throat, Nose, Anterior Nares, (Delawarr)	14.75	21.25	Freq: 20.0, 880.0
Female Genital System, Vulva, (Delawarr)	15.00	18.20	
Pulmonary Disorders, Bronchiectasis, (Delawarr)	15.00	22.50	
Large Intestine, Cecum, (Delawarr)	15.00	35.00	Freq. 10.0, 880.0
Female Genital System, Uterus, Ligaments, Two Broad Ligaments, (Delawarr)	15.00	62.00	
External & Physical Agent Disorders, Surgical Shock, (Delawarr)	15.50	19.75	
Female Genital System, Ovaries, Uterine Extremity, (Delawarr)	15.50	20.00	Freq. 1.0, 880.0
Mouth, Nose & Throat, Mouth, Tongue, Taste Buds, Sweetness, (Delawarr)	15.50	29.75	
Mouth, Nose & Throat, Pharynx, Epiglottis, Thin Triangular Lidlike Piece of Cartilage, (Delawarr)	15.50	43.50	that Folds Back Over the Opening of the Windpipe During Swallowing to Prevent Foreign Matter to Enter the Lungs
Breast, Mammary Glands, Veins, Circulus Venosus, (Delawarr)	15.60	68.25	
Central Nervous System, Brain, Fore, Pineal Gland, (Delawarr)	15.75	28.25	With Depression Check Need to Increase Melatonin, Noradrenaline, Serotonin Or Decrease Dopamine, Will Be Low for Nervous Conditions; Freq: 20.0, 5000.0
Heart, Valves, (Delawarr)	16.00	14.00	Freq: 20.0, 162.0
Empyema, Pus in a Body Cavity, (Delawarr)	16.00	14.50	
Mouth, Nose & Throat, Salivary Glands, Sublingual Gland, (Delawarr)	16.00	23.50	
Duodenum, (Delawarr)	16.00	24.50	Use 16.00-25.00 if Insect Stings, Migraines, Parathion, Poisons (Arsenic), D D T, Acidity in Burns, or Lindane is found in
Ductless Glands, Pineal Body, (Delawarr)	16.00	28.00	With Depression Check Need to Increase Melatonin, Noradrenaline, Serotonin Or Decrease Dopamine, Will Be Low for Nervous Conditions. Freq. 20.0, 5000.0
Mouth, Nose & Throat, Nose, Olfactory Region, (Delawarr)	16.25	17.00	Freq: 20.0, 880.0
Small Intestine, Muscular Coat, Longitudinal Fibers, (Delawarr)	16.25	24.00	Freq. 10.0, 880.0
Heart, Valves, Papillary Muscles, (Delawarr)	16.40	14.60	Freq: 20.0, 162.0, 240.0
Large Intestine, Descending Colon, (Delawarr)	16.40	15.00	Freq. 10.0, 880.0, 20.0
Large Intestine, Structure Of, Muscular Coat, Circular Fibers, (Delawarr)	16.40	18.00	Freq. 10.0, 880.0
Neurologic Disorders, Muscular Atrophies & Dystrophies, (Delawarr)	16.50	15.50	A Wasting Away of Body Tissue; or Failure to Develop
Plant, Corn, (Delawarr)	16.50	16.00	
Breast, Mammary Glands, Veins, (Delawarr)	16.60	22.75	
Blood, Pressure, High, (Delawarr)	16.75	26.00	Check Structure, Emotions, Hormones, Boron
Mouth, Nose & Throat, Pharynx, Tonsils, Lymphocytes, (Delawarr)	16.75	33.25	Freq: 20.0, 880.0
Mouth, Nose & Throat, Nose, Posterior Nares, (Delawarr)	17.00	16.75	Freq: 20.0, 880.0
Reticulo-Endothelial System, Cells Comprising Loose Reticulum In, Endothelium Lining, (Delawarr)	17.00	20.00	The Vasuclar Sinuses, Lymph Glands

Patterns of Information	Rate Left Right	Energetic Processes
Duodenum, Mucous Membrane, Circular Folds, (Delawarr)	17.20 28.00	
Small Intestine, Mucous Membrane, Villi, Capillaries, (Delawarr)	17.25 11.75	Freq. 10.0, 880.0
Large Intestine, Vermiform Appendix, Muscous Membrane, (Delawarr)	17.25 20.75	Freq. 10.0, 880.0
Mouth, Nose & Throat, Esophagus, (Delawarr)	17.25 29.25	
Rheumatism, Chronic, (Delawarr)	17.25 35.00	
External & Physical Agent Disorders, Electric Shock, (Delawarr)	17.50 15.00	
Arthritis (Delawarr)	17.50 16.00	Balance 4.00-7.00 For Stiffness, Balance also 10.00-9.00, 13.00-29.00, Adrenals, colon, Kidneys & Thyroid; Freq: 43.3
Arthritis, Rheumatoid (Delawarr)	17.50 19.00	Check For Mineral Deposits, Calcium, Carbon-Stop Massive Carbohydrate Intake, Silicon, Sodium-Use Sun-Dried Sea Salt, Uric Acid; Freq: 43.3
Stomach, Lymph Vessels, (Delawarr)	17.50 22.50	Freq: 20.0
Mouth, Nose & Throat, Pharynx, Tonsils, Lymphoid Tissue, (Delawarr)	17.50 28.00	Freq: 20.0, 880.0
Heart, Ventricular Septum, (Delawarr)	17.60 18.75	Freq: 20.0, 162.0
Breast, Mammary Glands, Gland Tissue, Lobes, (Delawarr)	17.75 11.75	
Bruises, (Delawarr)	17.75 18.50	
Breast, Mammary Glands, Fibrous Tissue, (Delawarr)	17.75 98.25	
Large Intestine, Transverse Colon, Transverse Mesocolon, (Delawarr)	17.90 16.20	Freq. 10.0, 880.0, 20.0
Large Intestine, Structure Of, Solitary Lymphatic Nodules, (Delawarr)	18.00 17.50	Freq. 10.0, 880.0
Small Intestine, Serous Coat, (Delawarr)	18.00 19.50	Freq. 10.0, 880.0
Liver, Right Lobe, (Delawarr)	18.00 29.00	
Skeletal, Dislocation, (Delawarr)	18.20 15.60	Use 3.00-80.00
Reticulo-Endothelial System, Cells Comprising Loose Reticulum In, Endothelium Lining, (Delawarr)	18.20 18.40	The Vascular Sinuses, Medulla Of Suprarenal Gland
Heart, Purkinje's System, (Delawarr)	18.20 22.60	Freq: 20.0, 162.0
Internal Ear, Membranous Labyrinth, Saccule, Macula of the Saccule, (Delawarr)	18.20 57.60	Freq: 20.0, 880.0
Otorhinolaryngological Disorders, Nose & Paranasal Sinus Disorders, Chronic Catarrh, (Delawarr)	18.25 18.30	Use Of 3.00-9.00 Adrenals is Effective. Freq. 43.3
Gall Bladder & Biliary Ducts, Gall Bladder, Nerves, of Hepatic Plexus, (Delawarr)	18.25 27.50	
Psychological Shock, Sexual, (Delawarr)	18.50 11.25	Power or Controlling Influence of One Person on Another That Can Not be Fought Off
Heart, Pericardium, Fibrous Sac, (Delawarr)	18.75 11.25	
Heart, Lymph Vessels, Deep Plexus, (Delawarr)	18.75 17.75	Freq: 20.0, 162.0
Mouth, Nose & Throat, Esophagus, Muscle, Fibers, (Delawarr)	18.75 24.50	Freq: 20.0, 240.0
Heart, Fossa Ovalis, (Delawarr)	18.90 16.50	Freq: 20.0, 162.0
Heart, Arteries, Left Coronary, Interventricular Branch, (Delawarr)	19.00 12.00	Freq: 20.0, 162.0
Urinary System, Bladder, Female, (Delawarr)	19.00 14.00	
Female Genital System, Uterus, Nerves, From Ovarian Plexus, (Delawarr)	19.00 18.50	
Ductless Glands, Pineal Body, (Delawarr)	19.00 45.75	Freq. 20.0, 5000.0
Mouth, Nose & Throat, Nose, (Delawarr)	19.00 96.00	Freq: 20.0, 880.0
Stomach, Gastric Glands, Fundus, Source of Hydrochloric Acid (HC1), (Delawarr)	19.20 30.00	Freq: 20.0
Psychological Shock, Emotional, (Delawarr)	19.25 19.50	
Synovial Effusion, (Delawarr)	19.25 36.00	
Ductless Glands, Suprarenal (Adrenal) Glands, (Delawarr)	19.25 44.75	
Contraction, (Delawarr)	19.25 62.00	

Patterns of Information	Rate Left	Rate Right	Energetic Processes
Mouth, Nose & Throat, Esophagus, Mucous Membrane, (Delawarr)	19.40	21.25	
Central Nervous System, Hind Brain, Cerebellum, Fourth Ventricle, (Delawarr)	19.50	10.00	With Depression Check Need To Decrease Enkephalins Area of Emotion
Psychological Shock, (Delawarr)	19.50	20.00	
Male Genital System, Prostate Gland, Nerves, (Delawarr)	19.50	95.00	Freq: 20.0, 2000.0
Psychological Shock, Mental Oppression, (Delawarr)	19.75	18.75	
Psychological Shock, Financial, (Delawarr)	19.75	20.00	
Breast, Mammary Glands, Arteries, From Internal Mammary Artery, (Delawarr)	19.75	22.00	
Stomach, Sub Mucous Coat, Lymphatics, (Delawarr)	19.75	36.50	Freq: 20.0
Large Intestine, Structure Of, Muscular Coat, Longitudinal Fibers, (Delawarr)	20.00	12.00	Freq. 10.0, 880.0
Gall Bladder & Biliary Ducts, Gall Bladder, Nerves, Right Phrenic Fibers, (Delawarr)	20.00	12.00	
Heart, Endocardium, Elastic Fibers, (Delawarr)	20.00	12.00	Freq: 20.0, 162.0
Female Genital System, Uterus, Ligaments, Posterior Ligament, (Delawarr)	20.00	17.50	
Heart, Myocardium, (Delawarr)	20.00	19.00	Freq: 20.0, 162.0
Mouth, Nose & Throat, Nose, Respiratory Region, (Delawarr)	20.00	20.75	Freq: 20.0, 880.0
Mouth, Nose & Throat, Esophagus, Stratified Epithelium, (Delawarr)	20.00	21.00	
Liver, Veins, (Delawarr)	20.00	23.75	
Mouth, Nose & Throat, Esophagus, Muscle Fibers, Circular, (Delawarr)	20.00	24.00	Freq: 20.0, 240.0
Mouth, Nose & Throat, Esophagus, (Delawarr)	20.00	43.00	
Female Genital System, Vagina, (Delawarr)	20.00	56.00	
Urinary System, Urethra, Spongy Erectile Tissue, (Delawarr)	20.25	19.75	
Pancreas, Nerves, From Vagus Nerve, (Delawarr)	20.25	26.00	Freq. 15.0
Pancreas, Arteries, From Splenic Artery, (Delawarr)	20.25	39.50	Freq. 15.0
External & Physical Agent Disorders, Burns, (Delawarr)	20.50	17.75	
Small Intestine, Ileum, Lowest Part, Opening to Large Intestine, (Delawarr)	20.50	22.00	Freq. 10.0, 880.0
Mouth, Nose & Throat, Nose, Respiratory Region, Lateral Wall, (Delawarr)	20.50	27.25	Freq: 20.0, 880.0
Mouth, Nose & Throat, Nose, Respiratory Region, Nasal Muscous Membrane, (Delawarr)	20.60	22.75	Freq: 20.0, 880.0
Mouth, Nose & Throat, Nose, Nasal Cavity, (Delawarr)	20.75	17.25	Freq: 20.0, 880.0
Mouth, Nose & Throat, Nose, Respiratory Region, Nasolacrimal Duct, (Delawarr)	20.75	19.50	Freq: 20.0, 880.0
Urinary System, Bladder, Nerves, Sympathetics Are Motor To Sphincter & Inhibitory To, (Delawarr)	20.75	29.50	Muscles Of Bladder Wall
Urinary System, Bladder, Male, (Delawarr)	21.00	15.00	
Small Intestine, Helpful Paired with Heart 2.00-76.00, (Delawarr)	21.00	36.00	Pair Color-Red; Emotion-Happiness; Sound-Laughter; Freq. 10.0, 880.0
Circulatory Disorders, Arteries, Arteriosclerosis, (Delawarr)	21.00	47.00	A Thickening of the Walls of the Arteries, Loss of Elasticity. Freq. 43.34
Mouth, Nose & Throat, Nose, Lower Nasal Cartilage, (Delawarr)	21.00	53.00	Freq: 20.0, 880.0
Large Intestine, Ileocecal Valve, (Delawarr)	21.00	69.00	Freq. 10.0, 880.0
Reticulo-Endothelial System, Motile Phagocytic Cells, Blood Macrophages, (Delawarr)	21.20	09.50	
Sinuses, Venous Sinuses, Dura Mater, Draining Blood From Brain, (Delawarr)	21.25	15.50	
Heart, Atrio-Ventricular Node, (Delawarr)	21.25	16.40	Freq: 20.0, 162.0

Patterns of Information	Rate Left	Rate Right	Energetic Processes
Mouth, Nose & Throat, Mouth, Arteries, (Delawarr)	21.25	19.50	
Urinary System, Bladder, Lymph Vessels, (Delawarr)	21.25	19.50	
Liver, Hepatic Artery, Conveying Oxygenated Blood To, Liver Cells, (Delawarr)	21.25	35.50	
Female Genital System, Vagina, (Delawarr)	21.25	54.50	
Central Nervous System, Hind Brain, Cerebellum, Superior Peduncle, (Delawarr)	21.50	21.25	With Depression Check Need To Decrease Enkephalins Area of Emotion
Stomach, Arteries, Gastroepiloic, Left, (Delawarr)	21.50	22.75	Freq: 20.0
Urinary System, Bladder, Mucous Coat, (Delawarr)	21.50	24.25	
Muscular Structure, Ligaments, (Delawarr)	21.50	45.50	
External & Physical Agent Disorders, Traumatic Shock, (Delawarr)	21.75	17.20	
Large Intestine, Structure Of, Muscular Coat, (Delawarr)	21.75	28.00	Freq. 10.0, 880.0
Pancreas, Tail of Pancreas, (Delawarr)	21.75	30.25	Freq. 15.0
Mouth, Nose & Throat, Pharynx, (Delawarr)	21.75	33.00	
Ulcers, Varicose, (Delawarr)	21.75	37.50	
Nerves, Impinged, (Delawarr)	21.80	23.50	
Viral Diseases, Respiratory Viral Diseases, Coryza, (Delawarr)	22.00	02.00	Color Lemon, spread by Coughs & Sneezes, Scan & Use 96.00-0.00
Muscular, Strain, (Delawarr)	22.00	10.75	
Pulmonary Disorders, Airways Obstructions, Bronchitis, (Delawarr)	22.00	17.00	
Urinary System, Bladder, Ligaments, (Delawarr)	22.00	19.40	
Throat, Pharynx, Adenoids, Adenoid Area, (Delawarr)	22.00	26.25	Freq: 20.0, 880.0
Heart, Lymph Vessels, (Delawarr)	22.00	27.40	Freq: 20.0, 162.0
Pancreas, Veins, (Delawarr)	22.00	31.75	Freq. 15.0
Exostosis, Bony Growth Arising from the Bone, (Delawarr)	22.00	32.00	
Central Nervous System, Cranial Nerves, Optic, (Delawarr)	22.00	82.00	
Mouth, Nose & Throat, Salivary Glands, Parotid Gland, Veins, (Delawarr)	22.20	30.25	
Concussion, (Delawarr)	22.25	16.50	
Duodenum, Mucous Membrane, Villi, (Delawarr)	22.25	17.00	
Duodenum, Nerves, Sympathetics, Inhibit Peristalsis, (Delawarr)	22.25	21.75	
Duodenum, Mucous Membrane, Villi, Goblet Cells, (Delawarr)	22.25	25.10	
Spleen & Thymus, Spleen, Internal Fibro-elastic Coat, (Delawarr)	22.25	26.50	Freq: 20.0
Male Genital System, Seminal Vesicles, Outer Fibrous Layer, (Delawarr)	22.25	48.50	
Heart, Nerves, Cardiac Plexus, Deep, (Delawarr)	22.50	17.20	Freq: 20.0, 162.0
Small Intestine, Lymph Vessels, (Delawarr)	22.50	35.50	Freq. 10.0, 880.0
Eyes, Eyeball, Fibrous Coat, Sclera, (Delawarr)	22.50	37.00	Freq: 20.0
Mouth, Nose & Throat, Nose, Respiratory Region, Arteries, (Delawarr)	22.60	23.50	Freq: 20.0, 162.0
Mouth, Nose & Throat, Salivary Glands, Submaxillary Gland, (Delawarr)	22.75	14.50	
Heart, Atria, Left, (Delawarr)	22.75	18.80	Freq: 20.0, 162.0
Ductless Glands, Parathyroid Gland, Connective Tissue, Neutrophil Cells, (Delawarr)	22.75	20.00	
Central Nervous System, Hind Brain, Cerebellum, Inferior Peduncle, (Delawarr)	22.75	21.00	With Depression Check Need To Decrease Enkephalins Area of Emotion
Duodenum, Lymph Vessels, (Delawarr)	22.75	21.25	
Small Intestine, Mucous Membrane, Villi, Columnar Epithelium, (Delawarr)	22.75	25.25	Freq. 10.0, 880.0
Urinary System, Urethra, Sphincter Urethrae, (Delawarr)	22.75	27.50	

Patterns of Information	Rate Left	Rate Right	Energetic Processes
Respiratory System, Bronchi, Lymph Vessels, Passing to Pretracheal Gland, (Delawarr)	22.75	29.25	
Breast, Mammary Glands, Gland Tissue, Lobules, (Delawarr)	22.75	73.00	
Pericardial Effusion, (Delawarr)	23.00	06.75	
Female Genital System, Fallopian Tubes, Right, (Delawarr)	23.00	17.00	
Female Genital System, Fallopian Tubes, Left, (Delawarr)	23.00	19.00	
Peritoneal Effusion, (Delawarr)	23.00	19.25	
Duodenum, Peritoneal Covering, (Delawarr)	23.00	21.75	
Small Intestine, Mucous Membrane, Villi, Lymph Vessels, (Delawarr)	23.00	28.50	Freq. 10.0, 880.0
Phlebitis, (Delawarr)	23.00	28.50	
Urinary System, Kidneys, (Delawarr)	23.00	58.00	Use for Arsenic, D D T, Earache, Edema, Flu, Meningococcus, Migraines, Nausea, Poisons, Polio, Scar Tissue, Sniffles; Freq. 8.0, 880.0
Rheumatism, Gonorrheal, (Delawarr)	23.25	18.00	Freq: 600.0, 700.0
Musculoskeletal & Connective Tissue Disorders, Osteoarthritis, (Delawarr)	23.25	23.25	Slow Progressive, Degenerative Joint Disease; Wearing Away of Cartilage in Joints; Freq: 43.3
Respiratory, Pleural Effusion, (Delawarr)	23.25	32.75	
Mouth, Nose & Throat, Nose, Respiratory Region, Nasal Septum, (Delawarr)	23.40	20.00	Freq: 20.0, 880.0
Central Nervous System, Brain Areas, Postrolandic Area, (Delawarr)	23.50	25.00	
Pancreas, Head of Pancreas, Anterior Surface, (Delawarr)	23.50	26.00	Freq. 15.0
Cramps, (Delawarr)	23.60	10.00	Freq. 26.0
Large Intestine, Vermiform Appendix, Sub Mucous Coat, (Delawarr)	23.60	19.20	Freq. 10.0, 880.0
Heart, Atria, Right, (Delawarr)	23.60	27.20	Freq: 20.0, 162.0
Large Intestine, Vermiform Appendix, Longitudinal Muscular Fibers, (Delawarr)	23.60	44.60	Freq. 10.0, 880.0
Breast, Mammary Glands, Arteries, (Delawarr)	23.75	07.50	
Large Intestine, Descending Colon, (Delawarr)	23.75	14.00	Freq. 10.0, 880.0, 20.0
Gall Bladder & Biliary Ducts, Gall Bladder, Serous Coat, (Delawarr)	23.75	14.00	
Stomach, Gastric Glands, (Delawarr)	23.75	23.25	Freq: 20.0
Mouth, Nose & Throat, Esophagus, Sub Mucous Coat, (Delawarr)	23.75	24.00	
Urinary System, Bladder, Arteries, (Delawarr)	23.75	24.50	
Stomach, Pylorus, (Delawarr)	23.75	24.50	Freq: 20.0
Blood, Blood Corpuscles, White Corpuscles, Monocytes, (Delawarr)	23.75	25.50	
Stomach, Sub Mucous Coat, (Delawarr)	23.75	29.75	Freq: 20.0
Mouth, Nose & Throat, Pharynx, (Delawarr)	24.00	07.00	
Heart, Arteries, Left Coronary, (Delawarr)	24.00	12.60	Freq: 20.0, 162.0
Mouth, Nose & Throat, Nose, Respiratory Region, Veins, (Delawarr)	24.00	21.50	Freq: 20.0, 880.0
Small Intestine, Mucous Membrane, Aggregated Lymphatic Nodules, Peyer's Patch, (Delawarr)	24.00	22.00	Found in the Ileum of Small Intestine, Important in releasing Antibodies & Absorption of Nutrients; Freq. 10.0, 880.0
Gall Bladder & Biliary Ducts, Gall Bladder, Nerves, Celiac Branches, (Delawarr)	24.00	27.20	
Eyes, Eyeball, Fibrous Coat, Cornea, (Delawarr)	24.00	33.50	Freq: 20.0
Lymphatic System, Right Lymphatic Duct, Tributaries, Right Suclavian Trunk, (Delawarr)	24.00	38.25	Freq. 10.0, 880.0
Breast, Mammary Glands, Arteries, From Intercostal Artery, (Delawarr)	24.10	19.50	
Stomach, Inner Mucous Coat, (Delawarr)	24.25	22.75	Freq: 20.0
Pancreas, Tail of Pancreas, Superior Border, (Delawarr)	24.25	24.00	Freq. 15.0
Pancreas, Lymph Vessels, (Delawarr)	24.25	37.50	Freq. 15.0
Gall Bladder & Biliary Ducts, Gall Bladder, Mucous Coat, (Delawarr)	24.40	23.80	

Patterns of Information	Rate Left	Rate Right	Energetic Processes
Lesion, (Delawarr)	24.50	05.00	
Female Genital System, Vagina, Posterior Wall, (Delawarr)	24.50	15.00	
Duodenum, Mucous Membrane, Villi, Artery Leaving Villus, (Delawarr)	24.50	26.25	
Breast, Mammary Glands, (Delawarr)	24.75	24.75	
Adhesions, Pleural (Delawarr)	24.75	28.50	Freq. 43.0
Skeletal System, Bone, (Delawarr)	25.00	22.00	
Pancreas, Structure, Islets of Langerhans, Beta Cells, (Delawarr)	25.00	26.00	Freq. 15.0
Lymphatic System, Lymph Vessels, Middle Coat, (Delawarr)	25.00	26.00	Freq. 10.0, 880.0
Duodenum, Mucous Membrane, Villi, Muscle Cells, (Delawarr)	25.00	27.00	Freq. 20.0, 240.0
Tumor, Benign, Cyst, (Delawarr)	25.00	28.00	Freq: 666.0, 690.0, 740.0, 1840.0, 1998.0, 2000.0
Effusion, Escape of Fluid into a Part, (Delawarr)	25.00	29.50	
Stomach, Arteries, (Delawarr)	25.00	31.00	Freq: 20.0
Eyes, Eyeball, Vascular Coat, Choroid, (Delawarr)	25.00	37.00	Freq: 20.0
Reticulo-Endothelial System, Cells Of Bone Marrow, Red Marrow, (Delawarr)	25.10	18.50	
Urinary System, Bladder, Nerves, Parasympathetics Are Motor To Bladder Wall &, (Delawarr)	25.20	26.75	Inhibitory to The Sphincter
Pancreas, Tail of Pancreas, Inferior Surface, (Delawarr)	25.20	28.75	Freq. 15.0
Lymphatics Disorder, Lymph Glands, Lymphadenoma, (Delawarr)	25.25	25.50	Freq. 10.0, 880.0
Congenital, Predisposition, (Delawarr)	25.25	33.00	
Pulmonary Disorders, Pleural Disorders, Pleurisy, Often With Sharp Pain, Relieved, (Delawarr)	25.25	41.75	by Pressure, Increased by Movement Breathing or Coughing
Mouth, Nose & Throat, Pharynx, Tonsils, Tonsillar Artery, (Delawarr)	25.40	15.75	Freq: 20.0, 880.0
Urinary System, Urethra, (Delawarr)	25.50	13.25	
Breast, Mammary Glands, Fatty Tissue, (Delawarr)	25.50	17.75	Use 12.50-31.50
Pancreas, Arteries, From Branch of Hepatic Artery, (Delawarr)	25.50	23.50	Freq. 15.0
Liver, Hepatic Artery, Conveying Oxygenated Blood To, Bile Capillaries, (Delawarr)	25.50	29.00	
Large Intestine, Structure Of, Arteries, Mesenteric Superior, Rectal Superior, (Delawarr)	25.50	31.00	Freq. 10.0, 880.0
Duodenum, Nerves, Myenteric Plexus, Vagal Origin, (Delawarr)	25.60	23.60	
Male Genital System, Scrotum, (Delawarr)	25.60	25.75	
Large Intestine, Structure Of, Arteries, Mesenteric Superior, Rectal Middle, (Delawarr)	25.75	26.50	Freq. 10.0, 880.0
Index, Delawarr	25.75	30.00	
Stomach, Veins, (Delawarr)	26.00	32.00	Freq: 20.0
Lymphatic System, Thoracic Duct, (Delawarr)	26.00	33.00	Freq. 10.0, 880.0
Sinuses, Air Sinuses, (Delawarr)	26.00	33.00	
Large Intestine, Transverse Colon, Left Colic Flexure, (Delawarr)	26.20	10.40	Freq. 10.0, 880.0, 20.0
Disorders, Heart, Cardiac Asthma, (Delawarr)	26.25	09.00	Freq. 43.3
Urinary System, Bladder, (Delawarr)	26.25	20.50	
Mouth, Nose & Throat, Pharynx, Constrictor Muscles, (Delawarr)	26.25	23.75	Freq: 20.0, 240.0
Mouth, Nose & Throat, Pharynx, Tonsils, Tonsillar Plexus, (Delawarr)	26.25	24.75	Freq: 20.0, 880.0
Tumor, Malignant, Sarcoma, A Form of Cancer Which Develops from Connective Tissue, (Delawarr)	26.25	26.50	Freq: 2008.0
Tumor, Malignant, Sarcoma, A Form of Cancer Which Develops from Connective Tissue, (Delawarr)	26.25	26.50	Freq: 666.0, 690.0, 740.0, 1840.0, 1998.0, 2000.0

Patterns of Information	Rate Left	Rate Right	Energetic Processes
Small Intestine, Mucous Membrane, Villi, Goblet Cells, (Delawarr)	26.25	27.75	Freq. 10.0, 880.0
Duodenum, Mucous Membrane, (Delawarr)	26.25	42.25	
Skin, Hair & Nails, Sweat Glands, (Delawarr)	26.40	39.75	
Small Intestine, Arteries, Mesentric Superior, (Delawarr)	26.40	45.90	Freq. 10.0, 880.0
Heart, Purkinje's Fibers, (Delawarr)	26.50	18.00	Freq: 20.0, 162.0
Muscular Structure, Muscular Tissue, Nerve Supply, Motor Nerves, End Plates, (Delawarr)	26.50	28.00	With Depression Check Need To Increase or Decrease Acetylcholine
Female Genital System, Fallopian Tubes, Infundibulum, (Delawarr)	26.50	33.50	
Small Intestine, Arteries, Jejunal, (Delawarr)	26.60	34.10	Freq. 10.0, 880.0
Male Genital System, Testes, Testicular Veins, (Delawarr)	26.75	12.75	Freq. 1.0, 880.0
Female Genital System, Vagina, Nerves, From Vaginal Plexuses, (Delawarr)	26.75	15.00	
Male Genital System, Testes, Tunica Albuginea, (Delawarr)	26.75	20.25	Freq. 1.0, 880.0
Urinary System, Ureters, Nerves, (Delawarr)	26.75	20.75	
Urinary System, Bladder, Muscular Coat, Internal Longitudinal Layer, (Delawarr)	26.75	23.00	
Liver, Surfaces, Inferior or Visceral Surface, Porta Impression, (Delawarr)	26.75	24.00	
Respiratory System, Pleurae, Arteries, (Delawarr)	26.75	43.00	
Dermatitis, (Delawarr)	26.75	50.50	
Small Intestine, Sub Mucous Coat, (Delawarr)	26.75	53.50	Freq. 10.0, 880.0
Eyes, Eyeball, Orbital Cavity, Rectus Superior, (Delawarr)	26.75	56.00	Freq: 20.0
Central Nervous System, Spinal Cord, Meninges, of Brain & Spinal Cord, (Delawarr)	27.00	09.00	
Eyes, Eyeball, Iris, Pupil, (Delawarr)	27.00	25.00	Freq: 20.0
Duodenum, Mucous Membrane, Villi, Capillaries, (Delawarr)	27.00	25.00	
Breast, Mammary Glands, Gland Tissue, (Delawarr)	27.00	26.00	
Liver, Surfaces, Inferior or Visceral Surface, Colic Impression, (Delawarr)	27.00	27.50	
Central Nervous System, Cranial Nerves - 11th Pair, Spinal Accessory, (Delawarr)	27.00	33.40	Nerves that supply the traezius & sternocleidomastoid muscles
Gall Bladder & Biliary Ducts, Gall Bladder, Fibromuscular Coat, (Delawarr)	27.00	37.75	
Tumor, Benign, Fibroma, (Delawarr)	27.25	11.00	Freq: 666.0, 690.0, 740.0, 1840.0, 1998.0, 2000.0
Male Genital System, Testes, Lymph Vessels, Pre-aortic Lymph Glands, (Delawarr)	27.25	14.75	Freq. 1.0, 880.0
Hernia, (Delawarr)	27.25	23.50	Use 59.00-177.00, 31.00-2.00, 5.00-91.00, Then 6.00-81.00, Then 77.00-41.00, Check Staph
Pancreas, Nerves, From Splanchine Nerves, (Delawarr)	27.25	25.25	Freq. 15.0
Urinary System, Ureters, Mucous Coat, (Delawarr)	27.25	25.50	
Pancreas, Pancreatic Duct, Accessory Pancreatic Duct, (Delawarr)	27.25	35.00	Freq. 15.0
Skeletal System, Bone, Blood Vessels, (Delawarr)	27.25	36.40	
Female Genital System, Uterus, Cervix Of Uterus, Neck of, (Delawarr)	27.40	37.40	
Pancreas, Pancreatic Duct, (Delawarr)	27.50	21.75	Freq. 15.0
Male Genital System, Testes, Nerves, (Delawarr)	27.50	27.60	Freq. 1.0, 880.0
Pancreas, Tail of Pancreas, Anterior Surface, (Delawarr)	27.50	30.00	Freq. 15.0
Male Genital System, Scrotum, Skin, (Delawarr)	27.50	35.50	
Sinuses, Air Sinuses, Sphenoid, (Delawarr)	27.75	18.75	
Heart, Pericardium, Lymph Glands, Mediastinal, (Delawarr)	27.75	19.50	Freq: 20.0, 162.0
Stomach, Nerves, (Delawarr)	27.75	21.00	Freq: 20.0
Liver, Right Lobe, (Delawarr)	27.75	30.00	

Patterns of Information	Rate Left	Rate Right	Energetic Processes
Pancreas, Pancreatic Duct, Ampulla Of Bile Duct, (Delawarr)	27.75	31.75	Freq. 15.0
Ductless Glands, Pituitary Body, (Delawarr)	28.00	09.75	
Central Nervous System, Hind Brain, Medulla Oblongata, (Delawarr)	28.00	13.00	
Respiratory System, Lungs, Intrapulmonary Bronchi, Alveoli, (Delawarr)	28.00	14.75	Freq. 9.0
Heart, Veins, Small Cardiac, (Delawarr)	28.00	16.10	Freq: 20.0, 162.0
Ductless Glands, Suprarenal (Adrenal) Glands, Medulla, (Delawarr)	28.00	23.00	
Tumor, Benign, (Delawarr)	28.00	27.50	Freq: 666.0, 690.0, 740.0, 1840.0, 1998.0, 2000.0
Reticulo-Endothelial System, Cells Comprising Loose Reticulum In, Connective Tissues, (Delawarr)	28.00	30.20	
Stomach, Arteries, Plexus Of Larger Capillaries, (Delawarr)	28.00	33.10	Freq: 20.0
Stenosis, Constriction or Narrowing of a Passage, (Delawarr)	28.00	47.25	
Sprain, Trauma to a Joint that Causes Pain, (Delawarr)	28.00	48.50	
Female Genital System, Vagina, Anterior Wall, (Delawarr)	28.10	06.00	
Blood, Pressure, Low, (Delawarr)	28.25	16.00	
Large Intestine, Pelvic Colon, Pelvic Mesocolon, (Delawarr)	28.25	20.50	Freq. 10.0, 880.0, 20.0
Large Intestine, Anus, (Delawarr)	28.25	24.00	May be Affected by DDE; Freq. 10.0, 880.0
Urinary System, Urethra, Mucous Membrane, (Delawarr)	28.25	31.60	
Ductless Glands, Pituitary Body, Pars Intermedia, (Delawarr)	28.25	33.20	
Duodenum, Mucous Membrane, Duodenal Glands, (Delawarr)	28.25	41.00	
Central Nervous System, Cranial Nerves - 01st Pair, Olfactory, (Delawarr)	28.40	28.50	from the epithelial lining of the nose, entirely sensory, sense of smell, many travel to the temporal lobe of the cerebral cortex
Sinuses, Air Sinuses, Ethmoid, (Delawarr)	28.40	47.00	
Male Genital System, Testes, Tunica Vasculosa, (Delawarr)	28.50	18.50	Freq. 1.0, 880.0
Circulatory Disorders, Arteries, Arteritis, Inflammatory Disorder of the Arteries, (Delawarr)	28.50	23.50	
Hyperinsulinism, (Delawarr)	28.50	31.75	
Pain, Relieved by Food & Pressure, Increased by Movement, Muscle Ruptured, (Delawarr)	28.75	23.50	
Psychiatric Disorders, Drug Addiction, Alcoholism, Alcoholic, Substance Abuse, (Delawarr)	28.75	26.00	
Reticulo-Endothelial System, Cells Of Bone Marrow, Yellow Marrow, (Delawarr)	28.75	28.50	No longer produces blood cells
Circulatory System, Arteries, Middle Coat, (Delawarr)	28.75	29.50	
Edema, Fluid or Water Retention, (Delawarr)	28.75	31.50	Use Of 3.00-9.00 Adrenals is Effective, Check Posterior Pituitary 56.00-23.00, Use 6.00-17.00, 3.00-82.00, 23.00-23.00
Large Intestine, Sigmoid Flexure, (Delawarr)	28.75	42.50	Freq. 10.0, 880.0
Liver, Left Lobe, Falciform Ligament, (Delawarr)	29.00	18.00	
Respiratory System, Bronchi, Lymph Vessels, Passing to Paratracheal Gland, (Delawarr)	29.00	18.25	
Stomach, Gastric Glands, Pyloric, Source of Pepsin, (Delawarr)	29.00	23.80	Freq: 20.0
Urinary System, Bladder, Serous Coat, (Delawarr)	29.00	26.00	
Urinary System, Urethra, Muscular Coat, (Delawarr)	29.00	28.25	
Reticulo-Endothelial System, Cells Comprising Loose Reticulum In, Endothelium Lining, (Delawarr)	29.00	30.00	The Vascular Sinuses, Liver, Kupffer's Cells

Patterns of Information	Rate Left	Rate Right	Energetic Processes
Liver, Hepatic Artery, Conveying Arterial Blood To, Bile Ductules, (Delawarr)	29.00	33.00	
Liver, Left Lobe, Coronary Ligament, (Delawarr)	29.00	38.00	
Gall Bladder & Biliary Ducts, Gall Bladder, Veins, (Delawarr)	29.25	38.25	
Male Genital System, Testes, Testicular Veins, Pampiniform Plexus, (Delawarr)	29.50	27.25	Freq. 1.0, 880.0
Ductless Glands, Thymus Gland, Concentric Corpuscles, (Delawarr)	29.50	31.50	
Reticulo-Endothelial System, Cells Of Bone Marrow, Red Marrow, Normoblast, (Delawarr)	29.50	52.25	
Pancreas, Head of Pancreas, Posterior Surface, (Delawarr)	29.60	32.25	Freq. 15.0
Pancreas, Nerves, Through Splenic Plexus, (Delawarr)	29.60	32.50	Freq. 15.0
Breast, Mammary Glands, Lymph Vessels, (Delawarr)	29.60	39.50	
Pancreas, Arteries, From Branch of Mesenteric Artery, (Delawarr)	29.75	23.25	Freq. 15.0
Mouth, Nose & Throat, Pharynx, Nerves, Of Sphenopalatine Ganglion, (Delawarr)	29.75	24.00	
Mouth, Nose & Throat, Pharynx, Uvula Palatina, (Delawarr)	29.75	24.25	
Respiratory System, Bronchi, Nerves, (Delawarr)	29.75	28.50	
Lymphatic System, Thoracic Duct, Cisterna Chyli, (Delawarr)	29.75	29.50	Freq. 10.0, 880.0
Duodenum, Nerves, Meissner's Plexus, Vagal Origin, (Delawarr)	29.75	33.75	
Large Intestine, Omentum, (Delawarr)	29.75	44.00	Freq. 10.0, 880.0
Stomach, Arteries, Gastric, Right, (Delawarr)	29.90	34.00	Freq: 20.0
Carcinoma, Toxins, Malignant Tumor, (Delawarr)	30.00	00.00	Freq. 666.0, 690.0, 740.0, 1840.0, 1998.0, 2000.0
Carcinoma, Toxins, Malignant Tumor, (Delawarr)	30.00	00.00	Freq. 2127.0, 43.3
Breast, Mammary Glands, Arteries, Thoracic Lateral, (Delawarr)	30.00	23.50	Same as Abundance
Urinary System, Bladder, Muscular Coat, Middle Circular Layer, (Delawarr)	30.00	24.50	
Carcinoma, Cells, Abnormal Growth, (Delawarr)	30.00	100.00	Freq. 2127.0, 43.3
Liver, Veins, Portal, (Delawarr)	30.20	36.25	Formed By The Veins Of The Splanchnic Area Conveying Its Blood Into The Liver
Gall Bladder & Biliary Ducts, Gall Bladder, Sphincter, (Delawarr)	30.25	27.00	
Male Genital System, Testes, Tunica Vaginalis, (Delawarr)	30.25	28.00	Freq. 1.0, 880.0
Central Nervous System, Mid Brain, (Delawarr)	30.25	30.50	With Depression Check Need to Decrease Dopamine, Main Area of Brain, Area of Emotions
Duodenum, Mucous Membrane, Villi, Lacteals, (Delawarr)	30.25	30.50	
Reticulo-Endothelial System, Cells Comprising Loose Reticulum In, Meninges, (Delawarr)	30.40	59.00	
Mouth, Nose & Throat, Mouth, Tongue, Taste Buds, Bitterness, (Delawarr)	30.50	25.00	
Lymphatic System, Right Lymphatic Duct, Tributaries, Right Jugular Trunk, (Delawarr)	30.50	36.00	Freq. 10.0, 880.0
Heart, Tendon Of The Infundibulum, (Delawarr)	30.75	06.40	Freq: 20.0, 162.0
Diseases, Bacterial, Cholera, Saint Vitus' Dance, Nervous System Disorder, (Delawarr)	30.75	19.50	Characterized by Jerking or Irregular Movements, Check out for Lingering Symptoms
Liver, Nerves, From Vagus Nerves, (Delawarr)	30.75	34.50	
Mouth, Nose & Throat, Salivary Glands, (Delawarr)	30.75	36.25	
Liver, Hepatic Artery, Conveying Oxygenated Blood To, Sinusoids, (Delawarr)	30.75	36.25	
Liver, Surfaces, Anterior Surface, (Delawarr)	30.75	43.25	
Male Genital System, Spermatic Cord, Cremasteric Fascia, (Delawarr)	30.75	47.60	

Patterns of Information	Rate Left	Rate Right	Energetic Processes
Blood, Hemostatic Agents, Thromboplastin, (Delawarr)	30.75	55.25	
Female Genital System, Vagina, Posterior Fornix, (Delawarr)	31.00	16.25	
Large Intestine, Structure Of, Serous Coat, (Delawarr)	31.00	18.50	Freq. 10.0, 880.0
Circulatory Disorders, Artery, Aneurysm, Often With Radiating Pain, (Delawarr)	31.00	35.50	Sac Formed by Local Enlargement of a Weakened Wall of the Artery
Internal Ear, Bony Labyrinth, Cochlea, Scala Vestibuli, (Delawarr)	31.00	41.50	Freq: 20.0, 880.0
Liver, Veins, Hepatic, (Delawarr)	31.00	41.50	3 vessels returning blood from the liver and discharging into the inferior vena cava
Stomach, Arteries, Gastroepiloic, Right, (Delawarr)	31.00	42.60	Freq: 20.0
Respiratory System, Lungs, Arteries, (Delawarr)	31.00	91.00	Freq. 9.0
Respiratory System, Lungs, Arteries, (Delawarr)	31.00	93.00	Freq. 9.0
Blood, Plasma, (Delawarr)	31.20	19.00	
Stomach, Omentum, (Delawarr)	31.25	22.00	Freq: 20.0
Lymphatic System, Thoracic Duct, Tributaries, Left Jugular Trunk, (Delawarr)	31.25	25.50	Freq. 10.0, 880.0
Reticulo-Endothelial System, Cells Of Bone Marrow, Yellow Marrow, Polymorphonuclears, (Delawarr)	31.25	27.60	Freq: 20.0, 880.0
Spleen & Thymus, Spleen, Veins, Tributaries, (Delawarr)	31.25	29.25	Freq: 20.0
Stomach, Gastric Lymphatic Nodules, (Delawarr)	31.25	33.60	Freq: 20.0
Stomach, Nerves, Sympathetics, from Lumbar Sympathetic Trunk, (Delawarr)	31.25	37.50	Freq: 20.0
Ductless Glands, Parathyroid Gland, Arteries, (Delawarr)	31.50	26.00	
Duodenum, Veins, (Delawarr)	31.50	31.25	
Mouth, Nose & Throat, Mouth, Lips, Top, (Delawarr)	31.50	34.50	
Respiratory System, Paranasal Sinuses, Sphenoidal, (Delawarr)	31.50	37.00	
Eyes, Eyeball, Vascular Coat, (Delawarr)	31.50	41.50	Freq: 20.0
Urinary System, Calyces Minor, (Delawarr)	31.60	42.75	
Heart, Arteries, Right Coronary, Interventricular Branch, (Delawarr)	31.75	17.00	Freq: 20.0, 162.0
Urinary System, Bladder, Sub-mucous Coat, (Delawarr)	31.75	22.00	
Female Genital System, Fallopian Tubes, External, Serous Coat, (Delawarr)	31.75	27.50	
Central Nervous System, Cranial Nerves, Spinal Or Peripheral Nerves, (Delawarr)	31.75	27.75	
Stomach, Muscular Coat, (Delawarr)	31.75	29.75	Freq: 20.0
Central Nervous System, Cranial Nerves, Spinal Or Peripheral Nerves, Cervical 8 Pairs, (Delawarr)	31.75	30.75	
Radiation, Actinobacillosis, (Delawarr)	31.75	32.50	Infection by bacilli caused by Radiation
Male Genital System, Spermatic Cord, Internal Spermatic Fascia, (Delawarr)	31.75	34.75	
Lymphatic System, Lymph Glands, Sub-occipital, Draining Back Of The Scalp, (Delawarr)	31.75	38.50	Freq. 10.0, 880.0
Respiratory System, Trachea, (Delawarr)	31.75	75.40	
Respiratory System, Larynx, (Delawarr)	32.00	16.00	
Eyes, Eyeball, (Delawarr)	32.00	18.00	Freq. 20.0
Respiratory System, Bronchi, Lymph Vessels, (Delawarr)	32.00	20.75	
Urinary System, Kidneys, (Delawarr)	32.00	23.00	Use for Arsenic, D D T, Earache, Edema, Flu, Meningococcus, Migraines, Nausea, Poisons, Polio, Scar Tissue, Sniffles; Freq. 8.0, 880.0
Pancreas, Structure, Islets of Langerhans, Alpha Cells, Produces Glucagon,, (Delawarr)	32.00	24.00	Balances Blood Sugar; Freq. 15.0
Ductless Glands, Suprarenal (Adrenal) Glands, Cortex, Zona Reticularis, (Delawarr)	32.00	25.00	

Patterns of Information	Rate Left	Rate Right	Energetic Processes
Ductless Glands, Pituitary Body, Anterior Lobe, Beta Cells, Produce Insulin, (Delawarr)	32.00	32.50	
Sinuses, Air Sinuses, Maxillary, Lower Lungs, (Delawarr)	32.00	33.00	Freq. 9.0
Ductless Glands, Thyroid Gland, Vesicles, (Delawarr)	32.00	39.00	Freq. 20.0, 160.0
Mouth, Nose & Throat, Nose, Nasal Cavity, (Delawarr)	32.00	48.00	Freq: 20.0, 880.0
Central Nervous System, Mid Brain, Cerebral Peduncles, Corpora Quadrigemina, (Delawarr)	32.10	32.50	
Toxins, Toxic State, (Delawarr)	32.20	21.00	
Pulmonary Disorders, Pneumonia, (Delawarr)	32.25	11.00	Freq: 776.0
Spleen & Thymus, Spleen, Internal Fibro-elastic Coat, Splenic Cells, (Delawarr)	32.25	31.50	Freq: 20.0
Mouth, Nose & Throat, Mouth, Tongue, Taste Buds, Sourness, (Delawarr)	32.25	36.50	
Fungus, Bacteria, Infection, Actinomycosis, Lumpy Jaw, (Delawarr)	32.25	37.50	Results in Bony Degeneration of Jaws, Abscesses in Lungs, Intestines, Prolonged Therapy with Penicillin G is effective; Freq: 787.0
Lymphatic System, Lymph Glands, Auricularr, Draining Area Of Ear, (Delawarr)	32.25	38.75	Freq: 20.0, 880.0, 10.0
Respiratory System, Larynx, Ligaments, (Delawarr)	32.25	39.00	
Innervation, Impaired, (Delawarr)	32.50	17.00	
Respiratory System, Paranasal Sinuses, Maxillary, (Delawarr)	32.50	20.00	
Central Nervous System, Brain Areas, Postrolandic Area, Leg, (Delawarr)	32.50	25.00	
Circulatory System, Arteries, Outer Coat, (Delawarr)	32.50	30.75	
Small Intestine, Sub Mucous Coat, Lymphatics, (Delawarr)	32.50	37.10	Freq. 10.0, 880.0
Blood, Blood Corpuscles, White Corpuscles, Lymphocytes, Basophils, (Delawarr)	32.60	56.75	
Emphysema, (Delawarr)	32.75	12.00	
Respiratory System, Paranasal Sinuses, (Delawarr)	32.75	27.70	
Mouth, Nose & Throat, Mouth, Tongue, Taste Buds, (Delawarr)	32.75	28.75	
Large Intestine, Structure Of, Arteries, Mesentric Inferior, Rectal Inferior, (Delawarr)	32.75	29.25	Freq. 10.0, 880.0
Male Genital System, Testes, Lymph Vessels, (Delawarr)	32.75	31.50	Freq. 1.0, 880.0
Mouth, Nose & Throat, Mouth, Lips, Bottom, (Delawarr)	32.75	34.75	
Circulatory System, Arterioles, Arteriole Capillaries, (Delawarr)	32.75	41.60	
Eyes, Eyeball, Vitreous Body, (Delawarr)	32.75	43.20	Freq: 20.0
Pancreas, Tail of Pancreas, Posterior Surface, (Delawarr)	32.80	22.50	Freq. 15.0
Liver, Nerves, From Right Phrenic, (Delawarr)	33.00	16.00	
Enlargement, (Delawarr)	33.00	34.00	
Skeletal System, Cartilage, (Delawarr)	33.00	36.00	
Ductless Glands, Suprarenal (Adrenal) Glands, Cortex, (Delawarr)	33.00	39.00	Helpful With Reducing Cancer Cell Disturbance, Use For Rheumatic Fever, Heart, Flu & Lagrippe
Female Genital System, Uterus, Body Of Uterus, Fundus, (Delawarr)	33.00	40.25	
Mouth, Nose & Throat, Pharynx, Nerves, (Delawarr)	33.00	40.75	
Eyes, Eyeball, Orbital Cavity, Rectus Oblique Inferior, (Delawarr)	33.00	43.20	Freq: 20.0
Skin, Hair & Nails, Corium, (Delawarr)	33.00	46.00	
Liver, Hepatic Artery, Conveying Arterial Blood To, Connective Tissue, (Delawarr)	33.00	46.10	
Respiratory System, Diaphragm, (Delawarr)	33.00	58.00	

Patterns of Information	Rate Left	Rate Right	Energetic Processes
Reticulo-Endothelial System, Cells Of Bone Marrow, Yellow Marrow, Metamyelocytes, (Delawarr)	33.20	26.50	
Eyes, Eyeball, Retina, Layer Of Rods & Cones, (Delawarr)	33.20	49.50	Freq: 20.0
Spasm, Involuntary Sudden Movement, (Delawarr)	33.25	03.50	
Male Genital System, Testes, Testicular Artery, (Delawarr)	33.25	26.75	Freq. 1.0, 880.0
Liver, Hepatic Artery, Conveying Oxygenated Blood To, Bile Ductules, (Delawarr)	33.25	30.25	
Lymphatic System, Lymph Vessels, External Coat, (Delawarr)	33.25	34.20	Freq. 10.0, 880.0
Ductless Glands, Suprarenal (Adrenal) Glands, Medulla, Chromaffin Cells, (Delawarr)	33.40	25.60	
Pancreas, Structure, Paranucleus, (Delawarr)	33.50	21.75	Freq. 15.0
Central Nervous System, Cranial Nerves, Spinal Or Peripheral Nerves, Thoracic 12, (Delawarr)	33.50	30.25	Pairs
Liver, Hepatic Artery, Conveying Arterial Blood To, Walls Of Portal Vein, (Delawarr)	33.50	34.00	
Central Nervous System, Cranial Nerves, (Delawarr)	33.50	35.75	
Mouth, Nose & Throat, Pharynx, Adenoids, Adenoid Area, (Delawarr)	33.50	46.50	Freq: 20.0, 880.0
Internal Ear, Auditory Nerve, Cochlear Nerve, (Delawarr)	33.50	63.75	Freq: 20.0, 880.0
Breast, Mammary Glands, Gland Tissue, Ducts, (Delawarr)	33.60	25.50	
Male Genital System, Penis, Root, Crus, (Delawarr)	33.60	45.00	
Skin, Hair & Nails, Hair Follicle, Outer Layer of Dermic Coat, (Delawarr)	33.60	46.00	
Urinary System, Kidneys, Nerves, (Delawarr)	33.75	22.75	Freq. 8.0, 880.0
Lymphatic System, Right Lymphatic Duct, Tributaries, Right Bronchomediastinal Trunk, (Delawarr)	33.75	27.25	Freq. 10.0, 880.0
Fibroid, Fibrositis, (Delawarr)	33.75	31.00	
Eyes, Eyeball, Eyelids, Lymph Glands, (Delawarr)	33.75	35.60	Freq. 20.0
Large Intestine, Mesentery, (Delawarr)	33.75	43.50	Freq. 10.0, 880.0
Spleen & Thymus, Spleen, Peritoneum, (Delawarr)	33.75	66.25	Freq: 20.0
Breast, Mammary Glands, Nerves, From 6th Thoracic, (Delawarr)	33.90	26.90	
Respiratory System, Paranasal Sinuses, Ethmoidal, (Delawarr)	33.90	43.00	
Stomach, Nerves, Sympathetics, from Celiac Plexus, (Delawarr)	33.90	51.60	Freq: 20.0
Reticulo-Endothelial System, Cells Of Bone Marrow, Red Marrow, Erythrocyte, (Delawarr)	34.00	26.50	
Skin, Hair & Nails, Corium, Nerves, (Delawarr)	34.00	28.25	
Middle Ear, Auditory Ossicles, Incus, Ligaments Of Incus, (Delawarr)	34.00	32.00	Freq: 20.0, 880.0
Eyes, Eyeball, Vitreous Body, Hyaloid Membrane, (Delawarr)	34.00	32.20	Freq: 20.0
Liver, Surfaces, Posterior Surface, (Delawarr)	34.00	38.75	
Stomach, Nerves, Sympathetics, from Plexuses Around Epiloic Artery, (Delawarr)	34.00	58.20	Freq: 20.0
Urinary System, Pelvis Of Ureter, (Delawarr)	34.25	33.25	
Male Genital System, Seminal Vesicles, Internal Mucous Coat, (Delawarr)	34.25	37.50	
Liver, Surfaces, Superior Surface, (Delawarr)	34.25	40.75	
Duodenum, Arteries, (Delawarr)	34.25	43.00	
Small Intestine, Nerves, (Delawarr)	34.25	44.50	Freq. 10.0, 880.0
Duodenum, Suspensory Muscle, (Delawarr)	34.25	52.25	Freq. 20.0, 240.0
Male Genital System, Testes, Epididymis, (Delawarr)	34.40	33.50	Freq. 1.0, 880.0

Patterns of Information	Rate Left	Rate Right	Energetic Processes
Skin, Hair & Nails, Alveoli, Sebum Cutaneum or Debris, (Delawarr)	34.40	41.00	
Urinary System, Urachus, (Delawarr)	34.50	34.25	
Respiratory System, Lungs, Intrapulmonary Bronchi, Mucous Membrane, (Delawarr)	34.50	34.25	Freq. 9.0
Respiratory System, Pleurae, Unstriped Muscle, (Delawarr)	34.50	34.50	Freq: 20.0, 240.0
Eyes, Eyeball, Eyelids, Superior Tarsus, (Delawarr)	34.50	39.00	Freq: 20.0
Ductless Glands, Parathyroid Gland, (Delawarr)	34.60	42.00	If High Tendency to Clot, Balance Thyroid with Parathyroid, Located Near Or Alongside the Thyroid gland, Secretes Hormone Important In Controlling Calcium-Phosphorus Balance
Liver, Surfaces, Inferior or Visceral Surface, Duodenal Impression, (Delawarr)	34.75	24.50	
Lymphatic System, Lymph, Vessels, (Delawarr)	34.75	27.60	Freq. 10.0, 880.0
Ductless Glands, Pineal Body, Fibrils, (Delawarr)	34.75	30.00	Freq. 20.0, 5000.0
Central Nervous System, Fore Brain, Thalamus, Internal Capsule, (Delawarr)	34.75	33.20	With Depression Check Need To Increase Norepinephrines In
Skeletal System, Vertebrae, (Delawarr)	34.75	34.25	See Spinal
Duodenum, Sub Mucous Coat, (Delawarr)	34.75	48.75	
Central Nervous System, Brain Areas, Sex Areas, Endocrine Center, (Delawarr)	34.75	52.00	
Duodenum, Nerves, from Splanchnic Nerve, (Delawarr)	35.00	21.75	
Liver, Surfaces, Inferior or Visceral Surface, Gastric Impression, (Delawarr)	35.00	26.75	
Lymphatic System, Lymph Glands, External Iliac, Draining The Groin, (Delawarr)	35.00	28.80	Freq. 10.0, 880.0
Blood, Blood Corpuscles, Reticulocyte, (Delawarr)	35.00	29.75	
Liver, Surfaces, Inferior or Visceral Surface, Renal Impression, (Delawarr)	35.00	31.50	
Ductless Glands, Suprarenal (Adrenal) Glands, Sympathetic Nerve Fibers, (Delawarr)	35.00	36.25	
Lymphatic System, (Delawarr)	35.00	37.50	Freq. 10.0, 880.0
Central Nervous System, Autonomic Nervous System, Sympathetic System, Ganglia, (Delawarr)	35.00	37.50	
Large Intestine, Ascending Colon, (Delawarr)	35.00	47.25	Freq. 10.0, 880.0, 20.0
Skeletal System, Bone Marrow, (Delawarr)	35.00	92.00	See Reticulo-Endothelial System
Urinary System, Bladder, Muscular Coat, External Longitudinal Layer, (Delawarr)	35.10	24.25	
Urinary System, Ureters, Muscular Coat, (Delawarr)	35.25	25.25	
Reticulo-Endothelial System, Cells Comprising Loose Reticulum In, Endothelium Lining, (Delawarr)	35.25	35.00	The Vascular Sinuses, Anterior Pituitary
Ductless Glands, Pineal Body, Connective Tissue Cells, (Delawarr)	35.25	35.50	Freq. 20.0, 5000.0
Male Genital System, Penis, Bulb of Penis, (Delawarr)	35.25	52.25	
Skeletal System, Bone, Periosteum, (Delawarr)	35.40	30.00	
Skin, Hair & Nails, Epidermis, Horny Zone, Clear Layer, (Delawarr)	35.40	61.00	
Central Nervous System, Cranial Nerves, Spinal Or Peripheral Nerves, Lumbar 5 Pairs, (Delawarr)	35.50	18.50	
Urinary System, Kidneys, Lymph Vessels, (Delawarr)	35.50	19.75	Freq. 8.0, 880.0
Small Intestine, Muscular Coat, Circular Fibers, (Delawarr)	35.50	30.00	Freq. 10.0, 880.0
Duodenum, Sub Mucous Coat, Nerves, (Delawarr)	35.50	37.40	
Stomach, Veins, Splenic, (Delawarr)	35.50	38.75	Freq: 20.0
Internal Ear, Bony Labyrinth, Cochlea, Scala Tympani, (Delawarr)	35.50	43.00	Freq: 20.0, 880.0
Small Intestine, Veins, (Delawarr)	35.50	44.50	Freq. 10.0, 880.0

Patterns of Information	Rate Left	Rate Right	Energetic Processes
Central Nervous System, Spinal Cord, Subdural Space, (Delawarr)	35.50	73.00	
Central Nervous System, Brain Areas, Postrolandic Area, Body-Face, (Delawarr)	35.75	14.50	
Respiratory System, Bronchi, Nerves, From Sympathetic Trunks, (Delawarr)	35.75	29.00	
Pancreas, Structure, Centro-acinar Cells, (Delawarr)	35.75	33.00	Freq. 15.0
Stomach, Nerves, Sympathetic, from Plexuses Around Gastric Artery, (Delawarr)	35.75	41.30	Freq: 20.0
Eyes, Eyeball, Orbital Cavity, (Delawarr)	35.75	45.75	Freq: 20.0
Lymphatic System, Lymph Glands, Lymph Nodes, (Delawarr)	36.00	17.00	Help Battle, Virus & Other Abnormal Cellular Tissue, Check congestion, Inflammation, Toxins & Bacteria
Liver, Left Lobe, Left Triangular Ligament, (Delawarr)	36.00	28.00	
Ductless Glands, Thyroid Gland, Capsule of Connective Tissue, (Delawarr)	36.00	29.00	Freq. 20.0, 160.0
Pancreas, Structure, Islets of Langerhans, (Delawarr)	36.00	35.00	Freq. 15.0
Pancreas, Structure, Alveoli, (Delawarr)	36.00	35.25	Freq. 15.0
Spleen & Thymus, Spleen, Arteries, Splenic, (Delawarr)	36.00	66.50	Freq: 20.0
Stomach, Veins, Mesenteric, Superior, (Delawarr)	36.20	42.40	Freq: 20.0
Lymphatic System, Lymph Glands, Intercostal, Draining The Chest Wall, (Delawarr)	36.25	27.20	Freq. 10.0, 880.0
Ductless Glands, Pituitary Body, Pars Intermedia, Blood Vessels, (Delawarr)	36.25	32.20	
Ductless Glands, Thyroid Gland, Right Lobe, (Delawarr)	36.25	38.50	Freq. 20.0, 160.0
Sinuses, Air Sinuses, Antrum, (Delawarr)	36.40	31.25	
Heart, Arteries, (Delawarr)	36.50	23.60	Freq: 20.0, 162.0
Thrombosis, Blood Clot, (Delawarr)	36.50	25.50	
Gall Bladder & Biliary Ducts, Gall Bladder, Arteries, (Delawarr)	36.50	27.50	
Skin, Hair & Nails, Corium, Ducts, (Delawarr)	36.50	42.50	
Spleen & Thymus, Spleen, Veins, (Delawarr)	36.50	43.10	
Duodenum, Sub Mucous Coat, Blood Vessels, (Delawarr)	36.50	55.50	
Valvular Incompetence, (Delawarr)	36.50	60.25	
Duodenum, Mucous Membrane, Villi, Lymph Vessels, (Delawarr)	36.60	27.00	
Male Genital System, Scrotum, Dartos Muscle, (Delawarr)	36.60	33.00	Freq: 20.0, 240.0
Skin, Hair & Nails, Hair Follicle, (Delawarr)	36.60	46.00	
Male Genital System, Vas Deferens, (Delawarr)	36.60	59.75	
Reticulo-Endothelial System, Cells Comprising Loose Reticulum In, the Blood, Large, (Delawarr)	36.75	18.50	Mononuclear Leucocytes
Ductless Glands, Thymus Gland, Arteries, (Delawarr)	36.75	29.50	
Respiratory System, Paranasal Sinuses, Frontal, (Delawarr)	36.75	37.00	
Liver, Right Lobe, Spigelius's Lobe, (Delawarr)	36.75	48.50	
Liver, Surfaces, Right Surface, (Delawarr)	36.75	52.00	
Respiratory System, Bronchi, Nerves, From Vagus Nerve, (Delawarr)	37.00	07.75	
Central Nervous System, Brain Areas, Association Area, Frontal, (Delawarr)	37.00	26.00	
Spleen & Thymus, Spleen, Arteries, Branches In Spleen, (Delawarr)	37.00	55.00	Freq: 20.0
Male Genital System, Testes, (Delawarr)	37.00	94.00	Freq. 1.0, 880.0
Lymphatic System, Lymph Glands, Bronchial, Draining The Lungs, (Delawarr)	37.20	41.20	Freq. 9.0, 10.0, 880.0
Male Genital System, Testes, (Delawarr)	37.25	35.50	Freq. 1.0, 880.0

Patterns of Information	Rate Left	Rate Right	Energetic Processes
Spleen & Thymus, Spleen, Nerves, From Celiac Plexus, (Delawarr)	37.25	42.50	Freq: 20.0
Blood, Blood Corpuscles, White Corpuscles, Polymorphonuclears, (Delawarr)	37.25	56.75	
Respiratory System, Diaphragm, Diaphragm, Muscles, (Delawarr)	37.25	71.50	Freq: 20.0, 240.0
Central Nervous System, Cranial Nerves - 06th Pair, Abducens, (Delawarr)	37.50	38.00	Involved With The Coordinated Contraction Of Extrinsic Muscles Of Both Eyes, Synchronization Of Nerve Impulses From 3Rd, 4Th & 5Th
Respiratory System, Lungs, Nerves, (Delawarr)	37.50	76.50	Freq. 9.0
Central Nervous System, Mid Brain, Cerebral Peduncles, Tegmentum, (Delawarr)	37.75	35.00	
Stomach, Gastric Glands, Cardiac, (Delawarr)	37.75	36.20	Freq: 20.0
Lymphatic System, Lymph Vessels, Internal Coat, (Delawarr)	37.75	36.50	Freq. 10.0, 880.0
Lymphatic System, Lymph Glands, Epitrochlear, Draining The Hands & Forearms, (Delawarr)	37.75	37.00	Freq. 10.0, 880.0
Small Intestine, Sub Mucous Coat, Nerves, (Delawarr)	37.75	38.50	Freq. 10.0, 880.0
Ductless Glands, Suprarenal (Adrenal) Glands, Medulla, (Delawarr)	37.75	43.75	
Pain, Increased by Movement, Rheumatism, Acute, (Delawarr)	37.75	44.00	
Muscular Structure, Muscular Tissue, Nerve Supply, Motor Nerves, (Delawarr)	37.75	58.50	
Urinary System, Kidneys, Veins, (Delawarr)	38.00	12.50	Freq. 8.0, 880.0
Central Nervous System, Brain Areas, Prerolandic Area, Eyes & Head, (Delawarr)	38.00	37.50	Freq. 20.0
Mouth, Nose & Throat, Pharynx, Stratified Epithelium, (Delawarr)	38.00	37.90	
Urinary System, Kidneys, Fibrous Capsule, (Delawarr)	38.00	72.50	Freq. 8.0, 880.0
Male Genital System, Spermatic Cord, (Delawarr)	38.10	36.50	
Duodenum, Sub Mucous Coat, Lymphatics, (Delawarr)	38.10	55.25	
Female Genital System, Vagina, Erectile Tissue, (Delawarr)	38.25	17.50	
Lymphatic System, Lymph, Fluid, (Delawarr)	38.25	37.50	Freq. 10.0, 880.0
Lymphatic System, Thoracic Duct, Tributaries, Left Subclavian Trunk, (Delawarr)	38.25	37.75	Freq. 10.0, 880.0
Skeletal System, Bone, Blood Vessels, Haversian Canals, (Delawarr)	38.25	40.00	
Respiratory System, Lungs, Intrapulmonary Bronchi, Terminal Bronchioles, (Delawarr)	38.50	13.50	Freq. 9.0
Internal Ear, Membranous Labyrinth, Duct Of Cochlea, Limbus Laminae Spiralis, (Delawarr)	38.50	28.50	Freq: 20.0, 880.0
Respiratory System, Larynx, Muscles, (Delawarr)	38.50	36.20	Freq: 20.0, 240.0
Lymphatic System, Thoracic Duct, Tributaries, Left Bronchomediastinal Trunk, (Delawarr)	38.50	37.00	Freq. 10.0, 880.0
Skin, Hair & Nails, Alveoli, (Delawarr)	38.50	45.00	
Internal Ear, Bony Labyrinth, Fibro-Serous Membrane Lining, (Delawarr)	38.50	49.00	Freq: 20.0, 880.0
Spleen & Thymus, Spleen, Internal Fibro-elastic Coat, Splenic Pulp, (Delawarr)	38.60	34.50	Freq: 20.0
Tumor, Benign, Fibroma, Uterine, (Delawarr)	38.75	27.00	Freq: 666.0, 690.0, 740.0, 1840.0, 1998.0, 2000.0
Lymphatic System, Lymph Glands, Popliteal, Draining The Legs, Body-Back of Thighs, (Delawarr)	38.75	35.00	Freq. 10.0, 880.0
Skeletal System, Vertebrae, Intervertebral Discs, (Delawarr)	38.75	40.25	
Female Genital System, Uterus, Cervix Of Uterus, Parametrium, (Delawarr)	38.75	43.20	Loose Connective Tissue Around the Uterus
Lymphatic System, Endothelium Lining to Glands, (Delawarr)	38.75	43.60	Freq. 10.0, 880.0

Patterns of Information	Rate Left	Rate Right	Energetic Processes
Mouth, Nose & Throat, Salivary Glands, Parotid Gland, Arteries, (Delawarr)	39.00	12.50	
Mouth, Nose & Throat, Salivary Glands, Parotid Gland, Jaws, (Delawarr)	39.00	24.00	
Urinary System, Bladder, Universal, Helpful Paired with Kidney 3.00-82.00, (Delawarr)	39.00	30.00	Pair Color-Black; Emotion-Fear; Sound-Groaning; Freq. 8.0, 880.0
Ductless Glands, Thyroid Gland, Isthmus, (Delawarr)	39.00	35.00	Freq. 20.0, 160.0
Eyes, Eyeball, Ciliary Body, Ciliaris Muscle, Alters Shape of Crystalline Lens In, (Delawarr)	39.00	41.50	Accommodation. Freq. 20.0, 240.0
Stomach, Inner Mucous Coat, Columnar Epithelium, (Delawarr)	39.00	42.50	Freq: 20.0
Circulatory System, Capillaries, Cells Of Capillary Wall, (Delawarr)	39.00	50.00	
Female Genital System, Uterus, Muscular Coat, (Delawarr)	39.20	18.25	
Stomach, Veins, Between Glands, (Delawarr)	39.20	46.25	Freq: 20.0
Central Nervous System, Mid Brain, Cerebral Peduncles, (Delawarr)	39.25	19.00	
Lymphatic System, Lymph Glands, Deep Cervical, Draining The Tongue, Lower Pharynx,, (Delawarr)	39.25	29.50	Larynx &Thyroid; Freq. 10.0, 880.0
Compression, (Delawarr)	39.25	30.50	
Hepatic & Biliary Disorders, Calculus, Biliary, (Delawarr)	39.25	38.00	Abnormal Stony Mass In Bile Tract
Central Nervous System, Cranial Nerves - 09th Pair, Glossopharyngeal, (Delawarr)	39.25	38.50	Supplies muscle of pharynx (swallowing), carry impulses that are sensitive to changes in level of blood oxygen & carbon dioxide in carotid sinus, neurons carry impulses from taste buds in back of tongue, membranes of pharynx, tonsils & mid-ear cavity
Urinary System, Kidneys, Renal Capsules, Glomerular Capsule, Bowman's Capsule, (Delawarr)	39.25	43.10	Freq. 8.0, 880.0
Respiratory System, Lungs, Nerves, Broncho-constrictors From Vagus Nerve, (Delawarr)	39.25	46.00	Freq. 9.0
Skin, Hair & Nails, Alveoli, Basement Membrane, (Delawarr)	39.25	49.50	
Central Nervous System, Brain Areas, Auditory Area, Temporal Lobe, (Delawarr)	39.25	65.50	
Female Genital System, Uterus, Pelvic Visceromotor Fibers, (Delawarr)	39.40	28.00	
Central Nervous System, Cranial Nerves - 10th Pair, Vagus Nerve, (Delawarr)	39.50	27.50	Arising in Medulla, Oblongata & Providing Parasympathetic Innervation to Larynx, Lungs, Heart, Esophagus & abdominal organs, regulates heart rate, depth of respiration, digestive process
Respiratory System, Lungs, Pulmonary Substance, (Delawarr)	39.50	30.75	Freq. 9.0
Central Nervous System, Mid Brain, Cerebral Peduncles, Basis Pedunculi, (Delawarr)	39.50	32.00	
Cartilage, Displaced, Pain, Increased by Movement, (Delawarr)	39.50	33.50	
Otorhinolaryngological Disorders, Nose & Paranasal Sinus Disorders, Acute Catarrh, (Delawarr)	39.50	33.50	Use Of 3.00-9.00 Adrenals is Effective. Freq. 43.3
Central Nervous System, Brain Areas, Sensory Area, Uncinate Gyrus, (Delawarr)	39.50	39.00	
Skeletal System, Cartilage, Hyaline Cartilage, (Delawarr)	39.50	42.25	
Lymphatic System, Lymph, Pericardial Lymph Fluid, (Delawarr)	39.50	43.00	Freq. 10.0, 880.0
Duodenum, Muscular Coat, Circular Fibers, (Delawarr)	39.50	50.25	
Stomach, Veins, Portal, (Delawarr)	39.50	59.00	Freq: 20.0
Circulatory System, Arteries, Nerves, From Sympathetic Ganglia, (Delawarr)	39.50	84.00	

Patterns of Information	Rate Left	Rate Right	Energetic Processes
Stomach, Muscular Coat, Longitudinal Fibers, (Delawarr)	39.60	31.50	Freq: 20.0
Skeletal System, Bone, Osteogenetic Tissue, Young Bones Only, (Delawarr)	39.60	44.10	
Respiratory System, Bronchi, Nerves, From Recurrent Laryngeal Nerves, (Delawarr)	39.75	31.25	
Skeletal System, Cartilage, White Fibrocartilage, (Delawarr)	39.75	42.00	
Central Nervous System, Cranial Nerves - 12th Pair, Hypoglossal, (Delawarr)	39.75	45.00	Located beneath the tongue & supply the muscles of the tongue
Spleen & Thymus, Spleen, Arteries, Arterioles, (Delawarr)	39.75	45.25	Freq: 20.0
Muscular Structure, Muscular Tissue, Nerve Supply, (Delawarr)	39.75	49.00	
Female Genital System, Vagina, Nerves, From Pelvic Visceromotor Fibers, (Delawarr)	39.75	89.10	
Congestion, (Delawarr)	40.00	08.60	Use 4.00-81.00, 4.00-8.00, 40.00-91.00
Pulmonary Disorders, Airways Obstructions, Bronchial Asthma, (Delawarr)	40.00	18.50	Freq. 43.3
Blood, Homeostatic Agents, Fibrin, (Delawarr)	40.00	32.40	A Whitish, filamentous Protein formed by the action of thrombin on fibrinogen, entangles red & white blood cells & platelets forming a clot
Liver, Lymph Vessels, Deep, (Delawarr)	40.00	38.50	
Hernia, (Delawarr)	40.00	41.00	Use 59.00-177.00, 31.00-2.00, 5.00-91.00, Then 6.00-81.00, Then 77.00-41.00, Check Staph
Eyes, Eyeball, Retina, Outer Plexiform Layer, (Delawarr)	40.00	43.10	Freq: 20.0
Skin, Hair & Nails, Alveoli, Epithelial Cells, (Delawarr)	40.00	44.50	
Stomach, Nerves, Parasympathetics, Plexuses in Muscular Coat, (Delawarr)	40.00	44.50	Freq: 20.0
Large Intestine, Vermiform Appendix, (Delawarr)	40.00	46.00	Balance For Intestinal Gas (Flatulence), Typhoid Fever, Exanthematous Viral & Respiratory Viral Diseases, Malaria, Thyroid Imbalances, Eczema; Freq. 10.0, 880.0
Central Nervous System, Brain Areas, Prerolandic Area, Leg, Body-Thighs, (Delawarr)	40.00	53.50	
Gastrointestinal Disorders, Colitis, Colon, Large Intestine, Inflammation, (Delawarr)	40.00	91.00	
Liver, Surfaces, Inferior or Visceral Surface, (Delawarr)	40.25	20.75	
Liver, Right Lobe, Quadrate Lobe, (Delawarr)	40.25	35.00	
Male Genital System, Seminal Vesicles, Nerves Derived From Pelvic Plexuses, (Delawarr)	40.25	37.00	
Ductless Glands, Parathyroid Gland, Connective Tissue, Principal Cells, (Delawarr)	40.25	38.50	
Internal Ear, Bony Labyrinth, Bony Labyrinth, Semicircular Canals, Horizontal Canal, (Delawarr)	40.25	43.25	Freq: 20.0, 880.0
Respiratory System, Larynx, Nerves, (Delawarr)	40.25	43.30	
Liver, Right Lobe, Quadrate Lobe, (Delawarr)	40.25	45.00	
Internal Ear, Bony Labyrinth, Cochlea, Helicotrema, (Delawarr)	40.25	52.80	Freq: 20.0, 880.0
Female Genital System, Uterus, Cervix Of Uterus, Internal Os, Mouth, Opening, (Delawarr)	40.25	69.75	
Liver, Surfaces, Fissure For Ligamentum Venosum, (Delawarr)	40.30	39.00	
Central Nervous System, Brain Areas, Postrolandic Area, Arm, (Delawarr)	40.40	35.40	
Central Nervous System, Autonomic Nervous System, (Delawarr)	40.40	42.50	
Central Nervous System, Spinal Cord, Coccyx Balance, (Delawarr)	40.50	22.00	
Central Nervous System, Cranial Nerves, Spinal Or Peripheral Nerves, Sacral 5 Pairs, (Delawarr)	40.50	28.25	

Patterns of Information	Rate Left	Rate Right	Energetic Processes
Central Nervous System, Fore Brain, Thalamus, To Improve Memory, Nerves, (Delawarr)	40.50	34.50	With Depression Check Need To Increase Enkephalins, Melatonin, Norepinephrines, Serotonin Or Decrease Dopamine, Use With Hippocampus 32.5-42 & With Acetylcholine Hormone 39.5-36.25
Lymphatic System, Lymph Glands, Celiac, Draining Stomach, Spleen, Pancreas, Liver, (Delawarr)	40.50	35.20	Freq. 10.0, 880.0
Skin, Hair & Nails, Corium, Connective Tissue, (Delawarr)	40.50	41.00	
Gall Bladder & Biliary Ducts, Biliary Ducts, Bile Duct, (Delawarr)	40.50	52.75	
Eyes, Eyeball, Retina, (Delawarr)	40.50	54.30	Freq: 20.0
Respiratory System, Mediastinum, (Delawarr)	40.50	62.50	
Urinary System, Body-Ureters, Right, (Delawarr)	40.75	12.50	
Respiratory System, Bronchi, Submucous Layer, (Delawarr)	40.75	24.50	
Urinary System, Urethra, Skene's Glands, (Delawarr)	40.75	29.25	
Breast, Mammary Glands, Nipple, (Delawarr)	40.75	29.50	
Breast, Mammary Glands, Nerves, From 5th Thoracic, (Delawarr)	40.75	30.75	
Circulatory System, Veins, Internal Coat, (Delawarr)	40.75	31.75	
Ductless Glands, Suprarenal (Adrenal) Glands, Cortex, Zona Fasciculata, (Delawarr)	40.75	35.00	
Internal Ear, Bony Labyrinth, Cochlea, Secondary Spiral Lamina, (Delawarr)	40.75	46.60	Freq: 20.0, 880.0
Stomach, Nerves, Parasympathetics, Greater Posterior Gastric Nerve, (Delawarr)	40.75	47.50	Freq: 20.0
Mouth, Nose & Throat, Nose, Nerves For Skin, (Delawarr)	40.75	50.60	Freq: 20.0, 880.0
Internal Ear, Membranous Labyrinth, Utricle, (Delawarr)	40.75	51.20	Freq: 20.0, 880.0
Lymphatic System, Lymph Glands, Axillary, Draining Upper Limbs & Breasts, (Delawarr)	41.00	35.75	Freq. 10.0, 880.0
Liver, Hepatic Artery, Conveying Oxygenated Blood To, Liver Cells, (Delawarr)	41.00	39.75	
Mouth, Nose & Throat, Mouth, Nerves, (Delawarr)	41.00	42.00	
Female Genital System, Ovaries, Ovum, (Delawarr)	41.00	46.50	Freq: 752.0, 1.0, 880.0
Respiratory System, Lungs, Veins, (Delawarr)	41.00	53.75	Freq. 9.0
Eyes, Eyeball, Aqueous Humour, (Delawarr)	41.00	56.00	Freq. 20.0
Reticulo-Endothelial System, Cells Of Bone Marrow, Red Marrow, Erythroblast, (Delawarr)	41.20	56.00	
Female Genital System, Uterus, Nerves, From Hypogastric Plexus, (Delawarr)	41.25	19.00	
Central Nervous System, Cranial Nerves, Spinal Or Peripheral Nerves, Coccygeal 1 Pair, (Delawarr)	41.25	32.50	
Spleen & Thymus, Ductless Glands, Thymus, (Delawarr)	41.25	33.00	Freq: 20.0
Duodenum, Nerves, from Vagus Nerve, (Delawarr)	41.25	36.60	
Central Nervous System, Brain Areas, Visual Area, Occipital, (Delawarr)	41.25	46.00	
Urinary System, Kidneys, Renal Capsules, Epithelial Lining, (Delawarr)	41.50	10.75	Freq. 8.0, 880.0
Sinuses, Air Sinuses, Mastoid, (Delawarr)	41.50	37.50	
Skeletal System, Bone, (Delawarr)	41.50	42.60	
Central Nervous System, Hind Brain, Cerebellum, Middle Peduncle, (Delawarr)	41.50	47.25	With Depression Check Need To Decrease Enkephalins Area of Emotion
Middle Ear, Auditory Ossicles, Incus, (Delawarr)	41.50	49.60	Freq: 20.0, 880.0
Gall Bladder & Biliary Ducts, Biliary Ducts, Bile Duct, Cystic Duct, (Delawarr)	41.50	51.50	
Internal Ear, Bony Labyrinth, Cochlea, Fenestra Cochleae, (Delawarr)	41.50	54.50	Freq: 20.0, 880.0
Mouth, Nose & Throat, Mouth, Tongue, Filliform Papillae, (Delawarr)	41.50	77.00	

Patterns of Information	Rate Left	Rate Right	Energetic Processes
Central Nervous System, Brain Areas, Association Area, Post Antral, (Delawarr)	41.60	29.60	With Depression Check Need To Increase Noradrenaline, Serotonin, Or Decrease Dopamine, Area Of Memory
Internal Ear, Membranous Labyrinth, (Delawarr)	41.60	36.60	Freq: 20.0, 880.0
Breast, Mammary Glands, Nerves, (Delawarr)	41.60	39.60	
Central Nervous System, Mid Brain, Cerebral Peduncles, Aqueduct Of Mid Brain, (Delawarr)	41.75	16.50	
Urinary System, Bladder, Muscular Coat, (Delawarr)	41.75	25.00	
Spleen & Thymus, Spleen, Internal Fibro-elastic Coat, Trabeculae, (Delawarr)	41.75	26.00	Freq: 20.0
Central Nervous System, Cranial Nerves - 08th Pair, Vestibulocochlear, (Delawarr)	41.75	29.50	Transmit impulses of hearing & equilibrium
Large Intestine, Transverse Colon, (Delawarr)	41.75	39.25	Freq. 10.0, 880.0, 20.0
Reticulo-Endothelial System, Cells Of Bone Marrow, Red Marrow, Myelocytes, (Delawarr)	41.75	39.50	
Stomach, Muscular Coat, Oblique Fibers, (Delawarr)	41.75	44.50	Freq: 20.0
Stomach, Nerves, Parasympathetics, From Vagus Nerve, (Delawarr)	41.75	47.10	Freq: 20.0
Respiratory System, Larynx, Membranes, (Delawarr)	42.00	28.00	
Skin, Hair & Nails, Alveoli, Larger Cells Containing Fat, (Delawarr)	42.00	36.50	
Skin, Hair & Nails, Epidermis, Germinative Zone, Prickle Cell Layer, (Delawarr)	42.00	37.40	
Diseases, Mycobacterial, Tuberculosis, (Delawarr)	42.00	41.50	Use Of 3.00-9.00 Adrenals is Effective. Freq. 800.0, 43.3
Stomach, Nerves, Parasympathetics, Greater Anterior Gastric Nerve, (Delawarr)	42.00	48.20	Freq: 20.0
Mouth, Nose & Throat, Mouth, Veins, (Delawarr)	42.00	49.25	
Respiratory System, Bronchi, Arteries, (Delawarr)	42.00	56.00	
Small Intestine, Mucous Membrane, Villi, (Delawarr)	42.10	59.10	Freq. 10.0, 880.0
Urinary System, Kidneys, Cortical Substance, (Delawarr)	42.25	37.50	Freq. 8.0, 880.0
Pancreas, Head of Pancreas, (Delawarr)	42.25	43.50	Freq. 15.0
Endocarditis, (Delawarr)	42.25	44.00	
Internal Ear, Membranous Labyrinth, Macula Of The Utricle, Filaments Auditory Nerve, (Delawarr)	42.25	57.50	Freq: 20.0, 880.0
Ductless Glands, Thymus Gland, Veins, (Delawarr)	42.50	32.00	
Female Genital System, Uterus, External Serous Coat, (Delawarr)	42.50	32.60	
Stomach, Arteries, Gastric, Left, (Delawarr)	42.50	39.00	Freq: 20.0
Liver, Bile Pigment, (Delawarr)	42.50	40.25	
Liver, Left Lobe, Right Triangular Ligament, (Delawarr)	42.50	48.25	
Reticulo-Endothelial System, Cells Comprising Loose Reticulum In, Endotheliocyte, (Delawarr)	42.50	50.25	Classmatocytes
Eyes, Eyeball, Retina, Inner Nuclear Layer, (Delawarr)	42.60	53.80	Freq: 20.0
Ductless Glands, Suprarenal (Adrenal) Glands, Medulla, Arteries, (Delawarr)	42.75	31.25	
Respiratory System, Pleurae, Nerves, From Phrenic Nerve, (Delawarr)	42.75	33.50	
Sinuses, Air Sinuses, Frontal, (Delawarr)	42.75	39.00	
Reticulo-Endothelial System, Cells Of Bone Marrow, Reticulo-Endothelium, (Delawarr)	42.75	39.00	
Female Genital System, Uterus, Body Of Uterus, (Delawarr)	42.75	40.50	
Central Nervous System, Brain Areas, Speech, Broca's Area, (Delawarr)	42.75	47.25	
Skin, Hair & Nails, Nails, Nail Root, (Delawarr)	42.75	59.00	
Skin, Hair & Nails, Corium, Connective Tissue, Reticular Layer, (Delawarr)	43.00	41.00	

Patterns of Information	Rate Left	Rate Right	Energetic Processes
Central Nervous System, Cranial Nerves - 05th Pair, Trigeminal, (Delawarr)	43.00	42.00	Come from the Pons, Major sensory nerves of face, contain the impulses for muscles for chewing including masseter & temporalis
Lymphatic System, Lymph Glands, (Delawarr)	43.00	45.25	Freq. 10.0, 880.0
Respiratory System, Paranasal Sinuses, Nasal Mucous Membrane, (Delawarr)	43.00	46.00	
Sinuses, Venous Sinuses, Dura Mater, Postero-Superior Group, Transverse, (Delawarr)	43.00	46.50	
Pancreas, Insulin-Free Part, (Delawarr)	43.00	53.00	Freq. 15.0
Central Nervous System, Brain Areas, Sex, Frigidity, (Delawarr)	43.00	54.00	
Eyes, Eyeball, Vascular Coat, Choroid, Vascular Lamina, (Delawarr)	43.00	57.00	Freq: 20.0
Liver, Left Lobe, (Delawarr)	43.20	28.00	
Ductless Glands, Suprarenal (Adrenal) Glands, Cortex, Zona Glomerulosa, (Delawarr)	43.20	31.00	
Stomach, Sub Mucous Coat, Blood Vessels, (Delawarr)	43.20	34.50	Freq: 20.0
Ductless Glands, Parathyroid Gland, Veins, (Delawarr)	43.20	38.00	
Duodenum, Muscular Coat, (Delawarr)	43.25	22.50	
Fatty Degeneration, (Delawarr)	43.25	33.75	
Reticulo-Endothelial System, Cells Of Bone Marrow, Yellow Marrow, Lymphocytes, (Delawarr)	43.25	34.00	
Central Nervous System, Mid Brain, Cerebral Peduncles, Red Nucleus, (Delawarr)	43.25	38.25	
Female Genital System, Uterus, (Delawarr)	43.25	39.75	
Sinuses, Venous Sinuses, Dura Mater, Antero-Inferior Group, Cavernous, (Delawarr)	43.25	47.20	
Urinary System, Kidneys, Renal Capsules, Tubules, (Delawarr)	43.25	57.50	Freq. 8.0, 880.0
Central Nervous System, Brain Areas, Sleep Center, (Delawarr)	43.50	43.50	
Eyes, Eyeball, Lacrimal Apparatus, Lacrimal Gland, (Delawarr)	43.50	56.60	Freq: 20.0
Central Nervous System, Cranial Nerves - 03rd Pair, Oculomotor, (Delawarr)	43.60	43.50	From midbrain & travels to the 6 extrinsic muscle or superior rectus, medial rectus, inferior oblique, These nerves supply levator palpebrae superioris muscles that raise the upper child
Internal Ear, Bony Labyrinth, Perilymph, (Delawarr)	43.75	43.00	Freq: 20.0, 880.0
Urinary System, Ureters, (Delawarr)	43.75	48.00	
Ductless Glands, Suprarenal (Adrenal) Glands, Medulla, Venous Sinusoids, (Delawarr)	44.00	30.25	
Spleen & Thymus, Spleen, Nerves, (Delawarr)	44.00	35.00	Freq: 20.0
Detox (3) Delawarr, Lymphatic System, Support & Stimulate All Subtle Fields of	44.00	41.00	Detox Support Program, Affected By Lyme Disease; Freq. 10.0, 880.0
Eyes, Eyeball, Fibrous Coat, Cornea, Substantia Propria, Corneal Spaces, (Delawarr)	44.00	51.00	Freq: 20.0
Eyes, Eyeball, Retina, Outer Nuclear Layer, (Delawarr)	44.00	59.00	Freq: 20.0
Stomach, Nerves, Sympathetic, from Left Phrenic Plexus, (Delawarr)	44.00	62.50	Freq: 20.0
Skin, Hair & Nails, Corium, Sebaceous Glands, (Delawarr)	44.10	42.00	
Female Genital System, Ovaries, Germinal Epithelium, (Delawarr)	44.10	44.30	Freq. 1.0, 880.0
Sinuses, Venous Sinuses, Dura Mater, Antero-Inferior Group, Intercavernous, (Delawarr)	44.20	48.40	
Internal Ear, Membranous Labyrinth, Macula Of The Utricle, (Delawarr)	44.20	54.75	Freq: 20.0, 880.0
Lymphatic System, Lymph Glands, Superficial Cervical, Draining External Ear & Neck, (Delawarr)	44.25	29.20	Freq. 10.0, 880.0
Liver, Nerves, From Celiac Plexus, (Delawarr)	44.25	39.00	
Coronary, Thrombosis, Blood Clot, (Delawarr)	44.25	44.50	

Patterns of Information	Rate Left	Rate Right	Energetic Processes
Musculoskeletal & Connective Tissue Disorders, Ulcers, (Delawarr)	44.50	17.00	
Central Nervous System, Cranial Nerves - 02nd Pair, Optic, (Delawarr)	44.50	18.50	Carry Impulses Associated With Vision, Entirely Sensory, Originates In Retina Of Eyes, Nerve Fibers Can Terminate In The Thalamus Or The Visual Cortex In Occipital Lobe
Gastrointestinal Disorders, Appendicitis, Often With Sharp Pain, Relieved by Food, (Delawarr)	44.50	22.00	
Central Nervous System, Mid Brain, Cerebral Peduncles, Substantia Nigra, (Delawarr)	44.50	26.50	With Depression Check Need To Increase Melatonin, Or Decrease Serotonin
Central Nervous System, Hind Brain, Cerebellum, Grey Matter, (Delawarr)	44.50	38.00	With Depression Check Need To Decrease Enkephalins Area of Emotion
Large Intestine, Vermiform Appendix, Serous Coat, (Delawarr)	44.50	41.00	Freq. 10.0, 880.0
Respiratory System, Paranasal Sinuses, Antrum, (Delawarr)	44.50	45.50	
Gall Bladder & Biliary Ducts, Gall Bladder, Nerves, (Delawarr)	44.50	47.50	
Central Nervous System, Spinal Cord, Anterior Horns, (Delawarr)	44.50	51.00	
Ductless Glands, Thymus Gland, Lobules, (Delawarr)	44.75	38.50	
Central Nervous System, Spinal Cord, Grey Matter, (Delawarr)	44.75	39.50	
Central Nervous System, Spinal Cord, Subarachnoid Space, (Delawarr)	44.75	40.20	
Male Genital System, Penis, Nerves, From 2nd Sacral, (Delawarr)	44.75	52.20	
Central Nervous System, Fore Brain, Thalamus, Caudata Nucleus, (Delawarr)	45.00	28.00	With Depression Check Need To Increase Melatonin, Norepinephrines, Serotonin
Urinary System, Body-Ureters, Left, (Delawarr)	45.00	28.25	
Lymphatic System, Lymph Glands, Deep Inguinal, Draining Deep Tissues Of Lower Limbs, (Delawarr)	45.00	29.00	Use For Parasites, Hookworm, Pinworm, Arbovirus, Arenavirus, Encephalitis, Bronchitis, Pulmonary Disorders, Fright, Emphysema, Emotional Shock; Freq. 10.0, 880.0
Liver, Left Lobe, (Delawarr)	45.00	29.00	Use For Allergy, Poisons, Cancer, Diabetes, Edema, Phlegm, Insect Stings, Undulant Fever, Protein Digestion, Constipation, Mumps, Leukemia
Liver, Lymph Vessels, Hepatic Artery, (Delawarr)	45.00	29.25	
Ductless Glands, Pituitary Body, Posterior Lobe, Cell Columns, (Delawarr)	45.00	35.75	
Lymphatic System, Lymph, Lymphocytes, Lymphoid Activity, Cloning Sites for New, (Delawarr)	45.00	37.00	Immune Cells, Blood, Red Cells, White Pulp; Freq. 10.0, 880.0
Small Intestine, Nerves, From Sympathetic, Inhibit Peristalsis, (Delawarr)	45.00	39.25	Freq. 10.0, 880.0
Stomach, Inner Mucous Coat, Peptic Glands, (Delawarr)	45.00	39.75	Freq: 20.0
Blood, Homeostatic Agents, Thrombin, (Delawarr)	45.00	55.40	
Cell Structure, Protoplasm, Cytoplasm, Chromophil Granules, (Delawarr)	45.00	60.50	
Stomach, Muscular Coat, Circular Fibers, (Delawarr)	45.25	43.10	Freq: 20.0
Stomach, Inner Mucous Coat, Gland Ducts, (Delawarr)	45.25	49.00	Freq: 20.0
Central Nervous System, Autonomic Nervous System, Sympathetic System, Plexuses, (Delawarr)	45.25	60.10	
Liver, Nerves, Sympathetic, (Delawarr)	45.50	25.00	
Internal Ear, Bony Labyrinth, Vestibule, (Delawarr)	45.50	40.40	Freq: 20.0, 880.0
Liver, Hepatic Artery, (Delawarr)	45.50	40.75	
Ductless Glands, Thyroid Gland, Left Lobe, (Delawarr)	45.50	44.00	Freq. 20.0, 160.0
Central Nervous System, Brain Areas, Visuo-Psychic Area, (Delawarr)	45.50	45.00	
Reticulo-Endothelial System, Cells Comprising Loose Reticulum In, the Blood, (Delawarr)	45.50	50.25	
Skin, Hair & Nails, Corium, Veins, (Delawarr)	45.50	53.00	

Patterns of Information	Rate Left	Rate Right	Energetic Processes
Middle Ear, Pharyngotympanic Tube, Auditory Tube, (Delawarr)	45.75	48.75	Freq: 20.0, 880.0
Mouth, Nose & Throat, Nose, Nerves For Muscles, Facial Nerve, (Delawarr)	45.75	51.00	Freq: 20.0, 880.0, 240.0
Sinuses, Venous Sinuses, Dura Mater, Postero-Superior Group, Sigmoid, (Delawarr)	45.90	40.25	
Muscular Structure, Muscular Tissue, (Delawarr)	46.00	36.50	
Stomach, Nerves, Parasympathetics, Plexuses in Sub Mucous Coat, (Delawarr)	46.00	37.50	Freq: 20.0
Eyes, Eyeball, Retina, Stratum Opticum, (Delawarr)	46.00	39.00	Freq: 20.0
Respiratory System, Bronchi, Mucous Membrane, (Delawarr)	46.00	46.00	
Eyes, Eyeball, Retina, Ganglionic Layer, (Delawarr)	46.00	50.50	Freq: 20.0
Urinary System, Kidneys, Renal Capsules, Glomerulus, (Delawarr)	46.00	65.25	Freq. 8.0, 880.0
Small Intestine, Nerves, From Myenteric Plexus, (Delawarr)	46.10	56.60	Freq. 10.0, 880.0
Skin, Hair & Nails, Nails, Matrix, (Delawarr)	46.20	55.50	
Ductless Glands, Pituitary Body, Anterior Lobe, Chromophobe Cells, (Delawarr)	46.25	43.40	
Respiratory System, Pleurae, Serous Coat, (Delawarr)	46.25	44.50	
Internal Ear, Bony Labyrinth, Cochlea, Spiral Canal, (Delawarr)	46.25	46.00	Freq: 20.0, 880.0
Ductless Glands, Suprarenal (Adrenal) Glands, Trabeculae, (Delawarr)	46.25	47.70	
Female Genital System, Vagina, Muscular Coat, (Delawarr)	46.50	39.00	
Eyes, Eyeball, Fibrous Coat, Cornea, Substantia Propria, (Delawarr)	46.50	44.50	Freq: 20.0
Eyes, Eyeball, Vascular Coat, Choroid, Choriocapillary Lamina, (Delawarr)	46.50	45.40	Freq: 20.0
Eyes, Eyeball, Ciliary Body, Ciliary Ring, (Delawarr)	46.50	47.00	Freq. 20.0
Sinuses, Venous Sinuses, Dura Mater, Antero-Inferior Group, Basilar Network, (Delawarr)	46.50	47.20	
Internal Ear, Membranous Labyrinth, Tectorial Membrane, (Delawarr)	46.50	49.50	Freq: 20.0, 880.0
Sinuses, Venous Sinuses, Dura Mater, Postero-Superior Group, Straight, (Delawarr)	46.50	53.00	
Internal Ear, Spiral Organ, Membranous Labyrinth, (Delawarr)	46.50	71.00	Freq: 20.0, 880.0
Internal Ear, Membranous Labyrinth, Rods Of Corti, Inner, (Delawarr)	46.50	81.75	Freq: 20.0, 880.0
Circulatory System, Arteries, Nerves, (Delawarr)	46.60	54.00	
Female Genital System, Ovaries, Stroma, (Delawarr)	46.75	23.00	Freq. 1.0, 880.0
Calculus, Stones, (Delawarr)	46.75	34.00	
Liver, Posterior Surface, Groove for Inferior Vena Cava, (Delawarr)	46.75	41.50	
Female Genital System, Uterus, Cervix Of Uterus, External Os, Mouth, Opening, (Delawarr)	46.75	43.50	
Central Nervous System, Spinal Cord, White Matter, Medullated Nerve Fibers, Anterior, (Delawarr)	46.75	44.00	Column
Middle Ear, Auditory Ossicles, Stapes, Ligaments Of Stapes, (Delawarr)	46.75	49.00	Freq: 20.0, 880.0
Stomach, Nerves, Sympathetic, from Left Splanchnic Nerve, (Delawarr)	46.75	55.75	Freq: 20.0
Small Intestine, Nerves, From Splanchnic Nerve, (Delawarr)	46.75	58.00	Freq. 10.0, 880.0
Internal Ear, Bony Labyrinth, Semicircular Canals, Ampullae, (Delawarr)	46.80	35.40	Freq: 20.0, 880.0

Patterns of Information	Rate Left	Rate Right	Energetic Processes
Urinary System, Ureters, Fibrous Coat, (Delawarr)	47.00	23.50	
Central Nervous System, Fore Brain, Thalamus, Lentiform Nucleus, (Delawarr)	47.00	31.75	With Depression Check Need To Decrease Enkephalins, Serotonin
Lymphatic System, Lymph Glands, Sternal, Draining Front of Chest Wall & The, (Delawarr)	47.00	35.50	Abdominal Wall; Freq. 10.0, 880.0
Skin, Hair & Nails, Hair Follicle, Henle's Layer, (Delawarr)	47.00	40.25	
Internal Ear, Bony Labyrinth, Semicircular Canals, Crus Commune, (Delawarr)	47.00	47.00	Freq: 20.0, 880.0
Sinuses, Venous Sinuses, Coronary, (Delawarr)	47.00	47.75	
Female Genital System, Ovaries, Membrane Granulosa, (Delawarr)	47.00	49.00	Freq. 1.0, 880.0
Lymphatic System, Lymph Glands, Sub-Maxillary, Draining Face & Floor of Mouth, (Delawarr)	47.00	68.60	Freq. 10.0, 880.0
Stomach, Inner Mucous Coat, Goblet Cells, (Delawarr)	47.25	32.50	Freq: 20.0
Ductless Glands, Thymus Gland, , Areolar Tissue, (Delawarr)	47.25	35.50	
Skin, Hair & Nails, Epidermis, Germinative Zone, (Delawarr)	47.25	46.00	
Pericarditis, (Delawarr)	47.25	46.25	
Eyes, Eyeball, Orbital Cavity, Rectus Lateral, (Delawarr)	47.25	48.00	Freq: 20.0
Respiratory System, Bronchi, Cartilage Rings, (Delawarr)	47.25	65.20	
Stomach, Serous Coat, (Delawarr)	47.50	28.25	Freq: 20.0
Female Genital System, Fallopian Tubes, Fimbriated End, (Delawarr)	47.50	32.25	
Liver, Right Lobe, Caudate Lobe, (Delawarr)	47.50	37.00	
Ductless Glands, Suprarenal (Adrenal) Glands, Cortex, (Delawarr)	47.50	40.25	
Female Genital System, Fallopian Tubes, Middle Muscular Coat, (Delawarr)	47.50	48.00	
Cell Structure, Protoplasm, Nucleus, Nuclear Membrane, (Delawarr)	47.50	49.00	
Central Nervous System, Brain Areas, Sex Areas, Function, (Delawarr)	47.50	51.00	
Cell Structure, Protoplasm, Nucleus, (Delawarr)	47.60	66.00	
Female Genital System, Uterus, Internal Mucous Coat, (Delawarr)	47.75	39.50	
Skeletal System, Bone, Blood Vessels, Lamellae, (Delawarr)	47.75	50.10	
Respiratory System, Diaphragm, (Delawarr)	47.75	53.50	
Ductless Glands, Parathyroid Gland, Nerves, (Delawarr)	47.75	71.60	
Internal Ear, Bony Labyrinth, Cochlea, Osseous Spiral Lamina, (Delawarr)	48.00	44.00	Freq: 20.0, 880.0
Internal Ear, Membranous Labyrinth, Endolymph, (Delawarr)	48.00	45.50	Freq: 20.0, 880.0
Eyes, Eyeball, Vascular Coat, Choroid, Basal Lamina, (Delawarr)	48.00	46.00	Freq: 20.0
Eyes, Eyeball, Retina, (Delawarr)	48.00	54.00	Freq: 20.0
Internal Ear, Membranous Labyrinth, Saccule, Canalis Reuniens, (Delawarr)	48.00	70.00	Freq: 20.0, 880.0
Urinary System, Bladder, (Delawarr)	48.00	75.00	
Eyes, Eyeball, Iris, (Delawarr)	48.20	56.50	Freq: 20.0
Middle Ear, Pharyngotympanic Tube, Auditory Tube, Bony Part, (Delawarr)	48.20	60.25	Freq: 20.0, 880.0
Central Nervous System, Brain Areas, Balance & Equilibrium, Cerebellar Cortex, (Delawarr)	48.25	43.00	
Gall Bladder & Biliary Ducts, Gall Bladder, Arteries, Cystic, (Delawarr)	48.25	45.50	

Patterns of Information	Rate Left	Rate Right	Energetic Processes
Female Genital System, Ovaries, Graafian Follicles, (Delawarr)	48.25	46.00	Freq. 1.0, 880.0
Central Nervous System, Spinal Cord, Extradural Space, (Delawarr)	48.25	47.75	
Eyes, Eyeball, Fibrous Coat, Cornea, Substantia Propria, Lamellae, (Delawarr)	48.25	49.50	Freq: 20.0
Central Nervous System, Spinal Cord, White Matter, Medullated Nerve Fibers, Lateral, (Delawarr)	48.50	38.00	Column
Skin, Hair & Nails, Corium, Connective Tissue, Papillary Layer, (Delawarr)	48.50	49.50	
Eyes, Eyeball, Retina, Arteries, (Delawarr)	48.50	50.25	Freq: 20.0
Skeletal System, Cartilage, Elastic Fibrocartilage, (Delawarr)	48.50	51.25	
Internal Ear, Membranous Labyrinth, Rods Of Corti, (Delawarr)	48.50	54.50	Freq: 20.0, 880.0
Stomach, Nerves, Sympathetic, from Thoracic Sympathetic Trunk, (Delawarr)	48.60	53.50	Freq: 20.0
Central Nervous System, Cranial Nerves - 04th Pair, Trochlear, (Delawarr)	48.75	34.20	Small Nerves that carry impulses to & from the superior oblique muscles, helps with the voluntary movements of the eyeball
Central Nervous System, Spinal Cord, Meninges, Arachnoid Mater, (Delawarr)	48.75	40.40	
Central Nervous System, Brain Areas, Prerolandic Area, Trunk, (Delawarr)	48.75	46.25	
Internal Ear, Bony Labyrinth, Semicircular Canals, (Delawarr)	48.75	50.50	Freq: 20.0, 880.0
Central Nervous System, Spinal Cord, Grey Matter, Nerve Fibers, (Delawarr)	48.75	68.50	
Lymphatic System, Lymph Glands, Mediastinal, Draining Heart, Pericardium & Thymus, (Delawarr)	48.80	29.80	Freq. 10.0, 880.0
Middle Ear, Auditory Ossicles, Articulations Of Ossicles, (Delawarr)	48.80	49.00	Freq: 20.0, 880.0
Respiratory System, Lungs, Nerves, Pulmonary Plexus Anterior, (Delawarr)	49.00	39.50	Freq. 9.0
Blood, Homeostatic Agents, Prothrombin, (Delawarr)	49.00	50.50	
Middle Ear, Auditory Ossicles, (Delawarr)	49.00	62.75	Freq: 20.0, 880.0
Female Genital System, Vagina, Veins, (Delawarr)	49.10	35.25	
Internal Ear, Membranous Labyrinth, Saccule, Saccus Endlymphatics, (Delawarr)	49.20	57.25	Freq: 20.0, 880.0
Tumor, Benign, Polypus, (Delawarr)	49.25	14.00	Freq: 666.0, 690.0, 740.0, 1840.0, 1998.0, 2000.0
Skin, Hair & Nails, Epithelial Cells, Epithelium, (Delawarr)	49.40	53.00	
Central Nervous System, Cranial Nerves - 07th Pair, 7th Facial Nerve, (Delawarr)	49.50	33.75	Contain sensory neurons from taste buds of the tongue & supply motor nerves of face & scalp, carry neurons to the tear glands & salivary glands
Sinuses, Venous Sinuses, Dura Mater, Antero-Inferior Group, Petrosal Inferior, (Delawarr)	49.50	49.00	
Small Intestine, Nerves, From Meissner's Plexus From Which Fibers Also Pass to, (Delawarr)	49.50	49.25	Mucous Membrane; Freq. 10.0, 880.0
Skin, Hair & Nails, Nails, Nail Plate, (Delawarr)	49.50	55.25	
Central Nervous System, Spinal Cord, White Matter, Medullated Nerve Fibers, Posterior, (Delawarr)	49.60	37.50	Column
Blood, Blood Corpuscles, Hemoglobin, (Delawarr)	49.75	22.75	
Internal Ear, Membranous Labyrinth, Duct Of Cochlea, Lagaena, (Delawarr)	49.75	53.40	Freq: 20.0, 880.0
Large Intestine, Veins, Nerves, Sympathetic, Superior Mesenteric Ganglion, (Delawarr)	49.75	59.50	Viscera in Mesenteric Superior are served by; Freq. 10.0, 880.0
Large Intestine, Rectum, (Delawarr)	50.00	12.00	Freq. 10.0, 880.0

Patterns of Information	Rate Left	Rate Right	Energetic Processes
Respiratory System, Lungs, Intrapulmonary Bronchi, Bronchial Muscles, (Delawarr)	50.00	35.00	Freq. 9.0, 20.0, 240.0
Internal Ear, Membranous Labyrinth, Saccule, Filament Auditory Nerve, (Delawarr)	50.00	36.00	Freq: 20.0, 880.0
Central Nervous System, Fore Brain, Corpus Callosum, (Delawarr)	50.00	38.00	With Depression Check Need To Increase Enkephalins, Norepinephrines, Serotonin Or Decrease Dopamine, Melatonin
Eyes, Eyeball, Lens, (Delawarr)	50.00	38.25	Freq: 20.0
Circulatory System, Veins, (Delawarr)	50.00	40.75	
Central Nervous System, Autonomic Nervous System, Parasympathetic System, (Delawarr)	50.00	42.00	Hormone, Histamine
Respiratory System, Larynx, Veins, (Delawarr)	50.00	43.25	
Internal Ear, Bony Labyrinth, Cochlea, (Delawarr)	50.00	43.75	Freq: 20.0, 880.0
Cell Structure, Protoplasm, Nucleus, chromosomes, Y-Chromosomes, (Delawarr)	50.00	44.50	
Cell Structure, Protoplasm, (Delawarr)	50.00	46.60	
Reticulo-Endothelial System, Cells Of Bone Marrow, Yellow Marrow, Lymphoblasts, (Delawarr)	50.25	29.00	
Eyes, Eyeball, Vascular Coat, Choroid, Capillaries, (Delawarr)	50.25	48.75	Freq: 20.0
Respiratory System, Bronchi, Veins, (Delawarr)	50.40	42.25	
Cell Structure, Protoplasm, Nucleus, Nucleolus, (Delawarr)	50.50	31.50	
Skin, Hair & Nails, Epidermis, Horny Zone, Granular Layer, (Delawarr)	50.50	33.00	
Female Genital System, Fallopian Tubes, Internal Mucous Coat, (Delawarr)	50.50	33.50	
Circulatory System, Veins, Middle Coat, (Delawarr)	50.50	66.50	
Eyes, Eyeball, Orbital Cavity, Orbital Muscles, (Delawarr)	50.60	46.25	Freq: 20.0, 240.0
Sinuses, Venous Sinuses, Dura Mater, Antero-Inferior Group, Middle Meningeal, (Delawarr)	50.60	58.00	
Large Intestine, Pelvic Colon, (Delawarr)	50.75	22.75	Freq. 10.0, 880.0, 20.0
Eyes, Eyeball, Fibrous Coat, Sclera, (Delawarr)	50.75	41.50	Freq: 20.0
Gall Bladder & Biliary Ducts, Biliary Ducts, Spiral Valve, (Delawarr)	50.75	47.50	
Skin, Hair & Nails, Corium, Lymph Vessels, (Delawarr)	50.75	52.20	
Breast, Mammary Glands, Nerves, From 4th Thoracic, (Delawarr)	50.75	52.50	
Internal Ear, Membranous Labyrinth, Duct Of Cochlea, Stria Vascularis, (Delawarr)	50.75	55.50	Freq: 20.0, 880.0
Eyes, Eyeball, Iris, Nerves, (Delawarr)	50.75	61.20	Freq: 20.0
Respiratory System, Lungs, Arteries, Pulmonary Artery, (Delawarr)	50.75	64.75	Freq. 9.0
Mouth, Nose & Throat, Mouth, Gums, Toothpaste Program (11) Heal/Prevent, (Delawarr)	51.00	25.00	
Eyes, Eyeball, Ciliary Body, Ciliary Processes, (Delawarr)	51.00	48.00	Freq. 20.0
Eyes, Eyeball, Iris, Nerves, Long Ciliary, (Delawarr)	51.00	49.50	Freq: 20.0
Mouth, Nose & Throat, Nose, (Delawarr)	51.00	53.00	Freq: 20.0, 880.0
Sinuses, Venous Sinuses, Dura Mater, Postero-Superior Group, Occipital, (Delawarr)	51.00	56.00	
Blood, Blood Corpuscles, Red Corpuscles, Erythrocyte, (Delawarr)	51.20	41.30	
Eyes, Eyeball, Eyelids, (Delawarr)	51.20	64.60	Freq. 20.0
Eyes, Eyeball, Fibrous Coat, Sclera, Ciliary Nerves, (Delawarr)	51.25	59.00	Freq: 20.0
Female Genital System, Vagina, Arteries, (Delawarr)	51.25	60.25	
Middle Ear, Pharyngotympanic Tube, Auditory Tube, Mucous Membrane, (Delawarr)	51.40	50.20	Freq: 20.0, 880.0

Patterns of Information	Rate Left	Rate Right	Energetic Processes
Eyes, Eyeball, Ciliary Body, (Delawarr)	51.40	56.00	Freq. 20.0
Gall Bladder & Biliary Ducts, Gall Bladder, Lymph Vessels, (Delawarr)	51.50	32.75	
Middle Ear, Pharyngotympanic Tube, Auditory Tube, Tube Tonsil, (Delawarr)	51.50	55.25	Freq: 20.0, 880.0
Blood, Blood Corpuscles, White Corpuscles, Polymorphonuclears, Oxyphils, (Delawarr)	51.60	28.00	
Small Intestine, Mucous Membrane, (Delawarr)	51.75	42.25	Note C#; Freq: 281.0
Respiratory System, Bronchi, Smooth Muscle, (Delawarr)	51.75	42.50	Freq: 20.0, 240.0
Liver, Lymph Vessels, Superficial, (Delawarr)	51.75	49.50	
Ductless Glands, Pituitary Body, Posterior Lobe, Neuroglia Cells, (Delawarr)	52.00	28.50	
Ductless Glands, Pituitary Body, Anterior Lobe, Alpha Cells, (Delawarr)	52.00	32.50	
Liver, Nerves, (Delawarr)	52.00	36.00	
Blood, Blood Corpuscles, White Corpuscles, Lymphocytes, (Delawarr)	52.25	49.00	
Urinary System, Calyx Of Kidney, Cuplike Part or Cavity, (Delawarr)	52.25	50.50	Freq. 8.0, 880.0
Blood, Blood Corpuscles, Lymph Corpuscles, (Delawarr)	52.30	54.25	
Eyes, Eyeball, Retina, Inner Plexiform Layer, (Delawarr)	52.40	53.00	Freq: 20.0
Central Nervous System, Brain Areas, Prerolandic Area, Body-Face, (Delawarr)	52.50	28.50	
Respiratory System, Lungs, Intercostal Muscles, External, (Delawarr)	52.50	39.50	Freq. 9.0, 20.0, 240.0
Central Nervous System, Fore Brain, Thalamus, (Delawarr)	52.50	40.00	With Depression Check Need To Increase Norepinephrines In
Respiratory System, Lungs, Intercostal Muscles, Internal, (Delawarr)	52.50	42.50	Freq. 9.0, 20.0, 240.0
Male Genital System, Penis, Nerves, From 4th Sacral, (Delawarr)	52.50	52.75	
Ductless Glands, Parathyroid Gland, Connective Tissue, (Delawarr)	52.75	36.50	
Reticulo-Endothelial System, Cells Of Bone Marrow, Yellow Marrow, Promyelocytes, (Delawarr)	52.75	40.75	
Skin, Hair & Nails, Nails, Lymphatics Of Nail Bed, (Delawarr)	52.75	48.25	
Male Genital System, Prostate Gland, (Delawarr)	52.75	82.50	Surrounds the Neck of the Bladder & Urethra In Males, It is Part Glandular & part Muscular, Gland Secretes a Thin Opalescent, Slightly Alkaline Fluid Which Is Part of Semen; Freq: 20.0, 2000.0
Chill, (Delawarr)	53.00	31.50	
Female Genital System, Fallopian Tubes, (Delawarr)	53.00	50.50	
Central Nervous System, Fore Brain, Hypothalamus, Nerves, (Delawarr)	53.00	50.50	
Cell Structure, Protoplasm, Cytoplasm, (Delawarr)	53.20	26.00	
Respiratory System, Pleurae, (Delawarr)	53.25	51.30	
Eyes, Eyeball, (Delawarr)	53.25	53.00	Freq. 20.0
Male Genital System, Penis, Body, Corpus Spongiosum Penis, (Delawarr)	53.25	62.00	
Respiratory System, Pleurae, Basement Membrane, (Delawarr)	53.25	72.00	
Urinary System, Kidneys, Arteries, (Delawarr)	53.50	33.50	Freq. 8.0, 880.0
Blood, Clot, Hematoma, (Delawarr)	53.50	41.50	Check Thyroid/Parathyroid Balance
Muscular Structure, Muscular Tissue, Muscle Sheath, (Delawarr)	53.50	53.25	Freq: 20.0, 240.0
Central Nervous System, Brain Areas, Association Area, (Delawarr)	53.50	80.75	

Patterns of Information	Rate Left	Rate Right	Energetic Processes
Urinary System, Kidneys, Uriniferous Tubules, (Delawarr)	53.75	19.75	Freq. 8.0, 880.0
Internal Ear, Auditory Nerve, Vestibular Ganglion, (Delawarr)	53.80	55.60	Freq: 20.0, 880.0
Respiratory System, Lungs, Sub-Serous Tissue, (Delawarr)	54.00	40.00	Freq. 9.0
Respiratory System, Lungs, Sub-Serous Tissue, (Delawarr)	54.00	50.00	Freq. 9.0
Central Nervous System, Hind Brain, Cerebellum, White Matter, (Delawarr)	54.00	58.75	With Depression Check Need To Decrease Enkephalins Area of Emotion
Liver, Surfaces, Posterior Surface, Suprarenal (Adrenal) Impression, (Delawarr)	54.25	38.50	
Internal Ear, Membraneous Labyrinth, Hair Cells, (Delawarr)	54.25	43.75	Freq: 20.0, 880.0
Male Genital System, Spermatic Cord, Nerves, (Delawarr)	54.25	44.50	
Sinuses, Venous Sinuses, Dura Mater, Postero-Superior Group, Sagittal Superior, (Delawarr)	54.25	58.00	
Cell Structure, Protoplasm, Cytoplasm, Mitochrondria, In Brown Fat Cells, (Delawarr)	54.50	39.00	
Internal Ear, Bony Labyrinth, Cochlea, Modiolus, (Delawarr)	54.50	44.75	Freq: 20.0, 880.0
Cell Structure, Protoplasm, Nucleus, Genes, (Delawarr)	54.50	56.20	
Skin, Hair & Nails, Nails, Nail Wall, (Delawarr)	54.75	59.50	
Eyes, Eyeball, Fibrous Coat, (Delawarr)	55.00	47.50	Freq: 20.0
Muscular Structure, Tendons, Sheaths, (Delawarr)	55.00	64.50	
Eyes, Eyeball, Fibrous Coat, Sclera, External Surface, (Delawarr)	55.50	47.50	Freq: 20.0
Cell Structure, Protoplasm, Nucleus, Chromosomes, (Delawarr)	55.50	52.00	
Middle Ear, Auditory Ossicles, Malleus, (Delawarr)	55.50	61.75	Freq: 20.0, 880.0
Internal Ear, Membraneous Labyrinth, Duct Of Cochlea, Vestibular Membrane, (Delawarr)	55.60	42.75	Freq: 20.0, 880.0
Blood, Blood Corpuscles, Plasma Cell, (Delawarr)	55.60	61.75	With Depression Check Need to Decrease Enkephalins, Melatonin, Serotonin
Eyes, Eyeball, Fibrous Coat, Cornea, Corneal Epithelium, (Delawarr)	55.75	46.75	Freq: 20.0
Eyes, Eyeball, Iris, Muscular Fibers, (Delawarr)	55.75	58.50	Freq: 20.0
Cell Structure, Protoplasm, Cytoplasm, Hyaloplasm, (Delawarr)	55.75	64.50	
Eyes, Eyeball, Eyelids, Conjunctiva, (Delawarr)	56.00	22.00	Freq. 20.0
Ductless Glands, Pituitary Body, Posterior Lobe, (Delawarr)	56.00	23.00	With Depression Check Need to Increase Enkephalins, Serotonin, Affected by Lyme Disease
Large Intestine, Veins, Nerves, Sympathetics, Part Of The Lumbar Plexus, (Delawarr)	56.00	48.25	Viscera in Mesenteric Inferior are Served by; Freq. 10.0, 880.0
Skin, Hair & Nails, Epidermis, Germinative Zone, Basal Cell Layer, (Delawarr)	56.00	49.50	
Perforation, (Delawarr)	56.00	53.00	
Eyes, Eyeball, Vitreous Body, Hyaloid Canal, (Delawarr)	56.00	55.25	Freq: 20.0
Respiratory System, Trachea, Mucous Membrane, (Delawarr)	56.25	49.75	
Eyes, Eyeball, Lacrimal Apparatus, Lacrimal Sac, (Delawarr)	56.40	61.60	Freq: 20.0
Male Genital System, Spermatic Cord, Arteries, (Delawarr)	56.50	50.25	
Eyes, Eyeball, Orbital Cavity, Rectus Oblique Superior, (Delawarr)	56.50	56.50	Freq: 20.0
Muscular Structure, Muscular Tissue, Unstriped Or Involuntary, (Delawarr)	56.50	61.00	
Eyes, Eyeball, Lens, Nucleus, (Delawarr)	56.50	68.50	Freq: 20.0

Patterns of Information	Rate Left	Rate Right	Energetic Processes
Small Intestine, Nerves, Parasympathetics, Augement Peristalsis & Secretion of, (Delawarr)	56.60	43.00	Gastric Juice; Freq. 10.0, 880.0
Mouth, Nose & Throat, Salivary Glands, Parotid Gland, Nerves, (Delawarr)	56.60	52.00	
Muscular Structure, Muscular Tissue, Muscle Fibers, (Delawarr)	56.75	61.50	Freq: 20.0, 240.0
Respiratory System, Larynx, Arteries, (Delawarr)	57.00	39.00	
Middle Ear, Auditory Ossicles, Stapes, (Delawarr)	57.00	44.40	Freq: 20.0, 880.0
Gall Bladder & Biliary Ducts, Biliary Ducts, Common Hepatic Duct, (Delawarr)	57.00	44.50	
Internal Ear, Bony Labyrinth, Semicircular Canals, Superior Semicircular Canal, (Delawarr)	57.00	48.40	Freq: 20.0, 880.0
Skin, Hair & Nails, Corium, Arteries, (Delawarr)	57.00	49.50	
Internal Ear, Bony Labyrinth, Semicircular Canals, Posterior Semicircular Canal, (Delawarr)	57.00	52.75	Freq: 20.0, 880.0
Central Nervous System, Autonomic Nervous System, Parasympathetic System, (Delawarr)	57.00	62.50	Nerve Fiber
Reticulo-Endothelial System, Cells Of Bone Marrow, Red Marrow, Reticulocyte, (Delawarr)	57.00	64.00	
Central Nervous System, Hind Brain, Medulla Oblongata, Olivary Nucleus, (Delawarr)	57.00	83.50	
Ductless Glands, Thyroid Gland, (Delawarr)	57.00	98.00	Freq. 20.0, 160.0
Tumor, Malignant, (Delawarr)	57.20	40.50	Freq: 666.0, 690.0, 740.0, 1840.0, 1998.0, 2000.0
Muscular Structure, Muscular Tissue, Muscle Fibers, Sarcolemma, (Delawarr)	57.25	22.25	Freq: 20.0, 240.0
Lymphatic System, Lymph Glands, Superficial Inguinal, Draining Buttocks, Anus,, (Delawarr)	57.25	38.50	Perineum, External Genitals & Nearly All The Superficial Lymph Of Lower Limbs; Freq. 10.0, 880.0
Central Nervous System, Hind Brain, Cerebellum, Dentate Nucleus, (Delawarr)	57.25	39.75	With Depression Check Need To Decrease Enkephalins Area of Emotion
Central Nervous System, Brain Areas, Coordination Of Movement, Cerebellar Cortex, (Delawarr)	57.25	40.75	
Male Genital System, Spermatic Cord, External Spermatic Fascia, (Delawarr)	57.25	55.00	
Eyes, Eyeball, Lens, Capsule, (Delawarr)	57.25	56.50	Freq: 20.0
Respiratory System, Pleurae, Pleural Cavity, (Delawarr)	57.25	62.25	
Mouth, Nose & Throat, Pharynx, Mucous Coat, (Delawarr)	57.40	33.00	
Male Genital System, Prostate Gland, Arteries, (Delawarr)	57.50	51.75	Freq: 20.0, 2000.0
Sinuses, Venous Sinuses, Dura Mater, Antero-Inferior Group, Petrosal Superior, (Delawarr)	57.50	53.00	
Internal Ear, Auditory Nerve, Spiral Ganglion, (Delawarr)	57.50	57.00	Freq: 20.0, 880.0
Muscular Structure, Tendons, (Delawarr)	57.50	69.00	
Respiratory System, Lungs, Nerves, From Vagus Nerve, (Delawarr)	57.60	69.50	Freq. 9.0
Skeletal System, Cartilage, Cellular Cartilage, (Delawarr)	57.75	46.50	
Central Nervous System, Hind Brain, Cerebellum, Nerves, Epileptogenic Zone, (Delawarr)	57.75	46.75	With Depression Check Need To Decrease Enkephalins Area of Emotion
Eyes, Eyeball, Iris, Nerves, From Occulomotor Nerve, (Delawarr)	57.75	68.20	Freq: 20.0
Central Nervous System, Fore Brain, (Delawarr)	57.75	87.50	With Depression Check Need to Increase Serotonin, Main Area of Brain
Male Genital System, Prostate Gland, (Delawarr)	58.00	26.00	Surrounds the Neck of the Bladder & Urethra In Males, It is Part Glandular & part Muscular, Gland Secretes a Thin Opalescent, Slightly Alkaline Fluid Which Is Part of Semen; Freq: 20.0, 2000.0
Internal Ear, Membraneous Labyrinth, Rods Of Corti, Outer, (Delawarr)	58.00	64.50	Freq: 20.0, 880.0
Central Nervous System, Spinal Cord, Meninges, of Brain & Spinal Cord, (Delawarr)	58.00	82.00	

Patterns of Information	Rate Left	Rate Right	Energetic Processes
Skin, Hair & Nails, Pigment, Melanin, (Delawarr)	58.20	53.20	
Muscular Structure, Muscular Tissue, Blood Supply, (Delawarr)	58.25	43.50	
Female Genital System, Ovaries, Corpus Luteum, (Delawarr)	58.25	45.75	Freq. 1.0, 880.0
Skin, Hair & Nails, Corium, Muscular Fibers, (Delawarr)	58.40	46.50	
Sinuses, Venous Sinuses, Dura Mater, Postero-Superior Group, Sagittal Inferior, (Delawarr)	58.40	50.20	
Circulatory System, Arteries, Nerves, From Plexuses, (Delawarr)	58.40	76.20	
Spleen & Thymus, Spleen, Lymph Vessels, (Delawarr)	58.50	38.90	Freq: 20.0
Muscular Structure, Muscular Tissue, Muscle Cells, (Delawarr)	58.50	51.00	Freq: 20.0, 240.0
Central Nervous System, Brain Areas, Prerolandic Area, Arm, (Delawarr)	58.50	52.50	
Blood, Blood Corpuscles, Platelets, (Delawarr)	58.75	48.10	With Depression Check Need To Increase Norepinephrins, Or Decrease Noradrenaline
Muscular Structure, Ligaments, (Delawarr)	58.75	65.00	
Skin, Hair & Nails, Cortex Of Hair, (Delawarr)	58.75	65.75	
Mouth, Nose & Throat, Mouth, Teeth, Toothpaste Program (12) Whiten/Vitalize, (Delawarr)	59.00	20.00	
Central Nervous System, Spinal Cord, White Matter, Medullated Nerve Fibers, (Delawarr)	59.00	39.00	
Blood, Blood Corpuscles, White Corpuscles, Leucocytes, (Delawarr)	59.00	42.00	
Respiratory System, Lungs, Pulmonary Substance, Lobules, (Delawarr)	59.00	42.50	Freq. 9.0
Internal Ear, Arteries of the Labyrinth, Internal Auditory Artery, (Delawarr)	59.00	54.50	Freq: 20.0, 880.0
Eyes, Eyeball, Retina, Macula Lutea, (Delawarr)	59.00	55.00	Freq: 20.0
Middle Ear, Pharyngotympanic Tube, Auditory Tube, Cartilagenous Part, (Delawarr)	59.00	55.20	Freq: 20.0, 880.0
Cell Structure, Protoplasm, Cytocentrum, Centrioles, (Delawarr)	59.00	57.10	
Small Intestine, Nerves, From Plexus around Superior Mesenteric Artery, (Delawarr)	59.00	59.00	Freq. 10.0, 880.0
Spleen & Thymus, Ductless Glands, Thymus, (Delawarr)	59.00	77.00	Trains or Modifies T-Cells which regulate the humoral immune system; Freq: 20.0
Breast, Mammary Glands, (Delawarr)	59.00	92.00	
Circulatory System, Capillaries, Cells Of Envelope, (Delawarr)	59.25	52.50	
Respiratory System, Larynx, The Nine Cartilages, (Delawarr)	59.50	33.25	
Lymphatic System, Lymph Glands, Internal Iliac, Draining The Pelvic Viscera, (Delawarr)	59.50	44.00	Freq. 10.0, 880.0
Central Nervous System, Spinal Cord, Meninges, Dura Mater, (Delawarr)	59.50	52.50	
Skin, Hair & Nails, Epidermis, Horny Zone, (Delawarr)	59.60	52.50	
Eyes, Eyeball, Orbital Cavity, Rectus Inferior, (Delawarr)	60.00	27.50	Freq: 20.0
Respiratory System, Lungs, Nerves, Pulmonary Plexus Posterior, (Delawarr)	60.00	44.00	Freq. 9.0
Internal Ear, Bony Labyrinth, Vestibule, Vestibular Crest, (Delawarr)	60.00	50.25	Freq: 20.0, 880.0
Internal Ear, Bony Labyrinth, Cochlea, Basilar Membrane, (Delawarr)	60.00	52.20	Freq: 20.0, 880.0
Central Nervous System, Brain Areas, Sensory Area, Taste, (Delawarr)	60.00	55.00	
Eyes, Eyeball, Fibrous Coat, Cornea, (Delawarr)	60.10	54.25	Freq: 20.0

Patterns of Information	Rate Left	Rate Right	Energetic Processes
Internal Ear, Membraneous Labyrinth, Duct Of Cochlea, Basilar Membrane, (Delawarr)	60.50	58.25	Freq: 20.0, 880.0
Eyes, Eyeball, Lacrimal Apparatus, Lacrimal Ducts, (Delawarr)	60.50	59.25	Freq: 20.0
Skin, Hair & Nails, Nails, (Delawarr)	60.50	72.50	
Female Genital System, Vagina, Labium, (Delawarr)	60.60	36.25	
Female Genital System, Vagina, Lymph Vessels, (Delawarr)	60.75	11.25	
Male Genital System, Penis, Nerves, Pudenal Branch, (Delawarr)	60.75	35.75	
Cell Structure, Protoplasm, Nucleus, Chromosomes, X-Chromosomes, (Delawarr)	60.75	56.00	Ovaries, Ovum, Genes Of Combined with
Small Intestine, Nerves, From Celiac Ganglion, (Delawarr)	60.75	70.50	Freq. 10.0, 880.0
Eyes, Eyeball, Vitreous Body, Suspensory Ligament Of Lens, (Delawarr)	61.00	55.20	Freq: 20.0
Internal Ear, Arteries of the Labyrinth, (Delawarr)	61.00	57.75	Freq: 20.0, 880.0
Circulatory System, Veins, External Coat, (Delawarr)	61.00	73.10	
Eyes, Eyeball, Lacrimal Apparatus, (Delawarr)	61.20	64.20	Freq: 20.0
Ductless Glands, Thyroid Gland, Pyramidal Lobe, (Delawarr)	61.50	25.75	Freq. 20.0, 160.0
Respiratory System, Larynx, (Delawarr)	61.50	47.75	
Eyes, Eyeball, Fibrous Coat, Cornea, Posterior Elastic Lamina, Pectinate Ligament Of, (Delawarr)	61.50	49.25	Iris; Freq: 20.0
Female Genital System, Uterus, Arteries, From Internal Iliac Artery, (Delawarr)	61.50	51.00	
Internal Ear, Bony Labyrinth, Pyramid Of The Vestibule, (Delawarr)	61.50	55.60	Freq: 20.0, 880.0
Central Nervous System, Fore Brain, Lateral Ventricles, (Delawarr)	61.50	62.00	
Blood, Hemostatic Agents, Tissue Fibrinogen, (Delawarr)	61.50	69.00	A Protein that helps form blood clots, High Elevations Indicate Hidden Risk for Heart Disease
Muscular Structure, Muscular Tissue, Striped Or Voluntary, (Delawarr)	61.50	75.00	
Small Intestine, Mucous Membrane, Circular Folds, (Delawarr)	61.75	36.20	Freq. 10.0, 880.0
Circulatory System, Arteries, Nerves, From Peripheral Nerves, (Delawarr)	61.75	55.25	
Disorders, Heart, Angina Pectoris, Conditon Marked by Recurrent Sharp, Radiating,, (Delawarr)	62.00	30.00	Referred Pain in the Chest & Left Arm; Use Also 20.00-41.00, Decrease in Blood Supply to Heart Muscle
Respiratory System, Bronchi, Muscular Tissue, (Delawarr)	62.00	39.50	
Urinary System, Kidneys, Medullary Substance, (Delawarr)	62.00	66.50	Freq. 8.0, 880.0
Middle Ear, Auditory Ossicles, Malleus Ligaments Of Malleus, (Delawarr)	62.10	47.20	Freq: 20.0, 880.0
Internal Ear, Auditory Nerve, (Delawarr)	62.20	52.50	Freq: 20.0, 880.0
Spleen & Thymus, Spleen, Lymph Vessels, Lymphatic Nodules, (Delawarr)	62.25	47.20	Freq: 20.0
Respiratory System, Lungs, Intrapulmonary Bronchi, Outer Fibrous Coat, (Delawarr)	62.25	58.75	Freq. 9.0
Female Genital System, Uterus, Veins, Ending In Uterine Plexuses, (Delawarr)	62.50	08.25	
Skin, Hair & Nails, Nails, Glomus, Important In Peripheral Circulation, (Delawarr)	62.50	64.75	
Male Genital System, Prostate Gland, Muscular Tissue, (Delawarr)	62.75	69.00	Freq: 20.0, 2000.0
Central Nervous System, Brain Areas, Prerolandic Area, (Delawarr)	63.00	48.25	
Internal Ear, Membraneous Labyrinth, Semicircular Ducts, (Delawarr)	63.00	52.60	Freq: 20.0, 880.0

Patterns of Information	Rate Left	Rate Right	Energetic Processes
Respiratory System, Bronchi, Bronchials, (Delawarr)	63.00	74.00	
Respiratory System, Lungs, Nerves, From Sympathics, (Delawarr)	63.50	58.50	Freq. 9.0
Duodenum, Muscular Coat, Longitudenal Fibers, (Delawarr)	63.50	60.20	
Female Genital System, Vagina, Perineal Body, (Delawarr)	63.75	46.75	
Eyes, Eyeball, Iris, Nerves, Ciliary Ganglion Branches, (Delawarr)	63.75	65.10	Freq: 20.0
Blood, Blood Corpuscles, Plasma Salts, (Delawarr)	63.75	67.50	With Depression Check Need To Increase Norepinephrins, Or Decrease Enkephalins, Melatonin, Serotonin
Male Genital System, Penis, Body, (Delawarr)	64.00	54.25	
Cell Structure, Protoplasm, Nucleus, Chromatin Granules, (Delawarr)	64.10	65.30	
Central Nervous System, Spinal Cord, Meninges, Cerebrospinal Fluid, Choroid Plexuses, (Delawarr)	64.25	45.50	With Depression Check Need To Increase Enkephalins, Melatonin, Norepinephrins, Serotonin Or Decrease Dopamine
Male Genital System, Penis, Arteries, (Delawarr)	64.25	58.00	
Cell Structure, Protoplasm, Cytocentrum, Attraction Sphere, (Delawarr)	64.25	73.00	
Ductless Glands, Parathyroid Gland, Lymph Vessels, (Delawarr)	64.50	53.50	
Muscular Structure, Fasciae, Deep Fascia, (Delawarr)	64.50	57.10	Fascia Tissue Surrounds Organs & Muscles & Helps Keep Their Shape; Emotions Settle in these Tissues Which Tends To Lock Up the Fascia
Skin, Hair & Nails, Hair Follicle, Hyaline Layer of Dermic Coat, (Delawarr)	64.50	63.20	
Blood, Hemostatic Agents, Thrombokinase, (Delawarr)	64.60	40.00	
Internal Ear, Auditory Nerve, Vestibular Nerve, (Delawarr)	64.60	59.90	Freq: 20.0, 880.0
Large Intestine, Structure Of, Arteries, Mesenteric Superior, Branching To Cecum,, (Delawarr)	64.75	57.50	Appendix, Ascending Colon, Transverse Colon (Right Hyprochondriac Region); Freq. Freq. 10.0, 880.0
Eyes, Eyeball, Vascular Coat, Choroid, Suprachoroid Lamina, (Delawarr)	64.75	65.30	Freq: 20.0
Blood, Blood Corpuscles, White Corpuscles, Macrophages, (Delawarr)	65.00	59.00	
Blood, Hemostatic Agents, Fibrinogen, (Delawarr)	65.00	64.50	A Protein that helps form blood clots, High Elevations Indicate Hidden Risk for Heart Disease
Reticulo-Endothelial System, Cells Of Bone Marrow, Yellow Marrow, Myelocytes, (Delawarr)	65.25	48.25	
Central Nervous System, Fore Brain, Third Ventricle, (Delawarr)	65.25	58.00	
Skin, Hair & Nails, Nails, Nail Bed, (Delawarr)	65.50	55.10	
Blood, Hemostatic Agents, Heparin, (Delawarr)	65.75	59.00	
Respiratory System, Larynx, Mucous Membrane, (Delawarr)	66.00	49.50	
Eyes, Eyeball, Iris, Stroma, (Delawarr)	66.20	55.50	Freq: 20.0
Ductless Glands, Pituitary Body, Posterior Lobe, (Delawarr)	66.25	55.75	With Depression Check Need to Increase Enkephalins, Serotonin, Affected by Lyme Disease
Muscular Structure, Muscular Tissue, Nerve Supply, Sensory Nerves, Muscle Spindles, (Delawarr)	66.60	63.00	Freq: 20.0, 240.0
Urinary System, Bladder, Veins, (Delawarr)	66.75	29.60	
Circulatory System, Capillaries, (Delawarr)	66.75	44.25	
Diet, Improper, (Delawarr)	67.00	27.00	
Lymphatic System, Lymph Glands, Lumbar, Draining Kidneys, Adrenals, Ovaries,, (Delawarr)	67.00	38.50	& Testicles. Freq. 1.0, 880.0, 10.0
Circulatory System, Arteries, Internal Coat, (Delawarr)	67.00	59.50	
Skeletal System, Bone, Blood Vessels, Nutrient Artery, (Delawarr)	67.00	61.00	
Reticulo-Endothelial System, Cells Of Bone Marrow, Yellow Marrow, Myeloblasts, (Delawarr)	67.25	42.50	

Patterns of Information	Rate Left	Rate Right	Energetic Processes
Internal Ear, Membraneous Labyrinth, Semicircular Ducts, Ampullae, (Delawarr)	67.75	45.75	Freq: 20.0, 880.0
Circulatory System, Arteries, (Delawarr)	68.00	71.00	
Male Genital System, Seminal Vesicles, Muscular Coat, (Delawarr)	68.25	53.50	
Central Nervous System, Spinal Cord, Meninges, Cerebrospinal Fluid, (Delawarr)	68.25	57.50	With Depression Check Need To Increase Enkephalins, Noradrenaline, Norepinephrins, Serotonin Or Decrease Dopamine, Melatonin
Male Genital System, Testes, Nerves, From Aortic Plexus, (Delawarr)	68.25	57.50	Freq. 1.0, 880.0
Male Genital System, Penis, Nerves, (Delawarr)	68.25	59.00	
Reticulo-Endothelial System, Cells Of Bone Marrow, Yellow Marrow, Monoblasts, (Delawarr)	68.25	80.50	
Internal Ear, Bony Labyrinth, (Delawarr)	68.50	46.00	Freq: 20.0, 880.0
Muscular Structure, Fasciae, Synovial Sheaths, (Delawarr)	68.50	61.30	
Internal Ear, Wax Or Cerumen, (Delawarr)	68.75	75.00	Freq: 20.0, 880.0
Mouth, Nose & Throat, Pharynx, Eustachian Tubes, (Delawarr)	69.50	38.00	
Blood, Blood Corpuscles, White Corpuscles, Polymorphonuclears, Neutrophils, (Delawarr)	69.50	69.00	
Central Nervous System, Brain Areas, Sex Areas. Frustration, (Delawarr)	69.50	71.00	Base, Anterior Auditory Artery
Cell Structure, Protoplasm, Cytoplasm, Tonofibrils, (Delawarr)	69.75	55.50	
Skin, Hair & Nails, Epidermis, (Delawarr)	70.00	02.50	
Eyes, Eyeball, Vascular Coat, Choroid, (Delawarr)	70.00	55.50	Freq: 20.0
Blood, Blood Corpuscles, White Corpuscles, Eosonophils, (Delawarr)	70.00	75.00	If higher than GV an indicator of parasite or allergies
Internal Ear, Membraneous Labyrinth, Semicircular Ducts, Crista Ampullaris, (Delawarr)	70.10	60.60	Freq: 20.0, 880.0
Circulatory System, Arteries, Nerves, From Perivascular Plexuses, Splanchnic Arteries, (Delawarr)	70.25	55.25	
Central Nervous System, Spinal Cord, White Matter, (Delawarr)	70.50	28.75	
Internal Ear, Membraneous Labyrinth, Duct of Cochlea, (Delawarr)	70.50	60.25	Freq: 20.0, 880.0
Large Intestine, Veins, Nerves, Sympathetics, Celiac Ganglion, (Delawarr)	70.50	77.75	Viscera in Mesenteric Superior are served by; Freq. 10.0, 880.0
Male Genital System, Penis, Root, (Delawarr)	71.25	54.25	
Reticulo-Endothelial System, Cells Of Bone Marrow, Reticulo-Endothelium, Plasma, (Delawarr)	71.30	59.00	Cells, Activated B Cell
Small Intestine, Nerves, From Vagus Nerve, (Delawarr)	71.40	72.60	Freq. 10.0, 880.0
Circulatory System, Veins, Valves, (Delawarr)	71.50	57.20	
Central Nervous System, Spinal Cord, Grey Matter, Neuroglia, (Delawarr)	71.50	71.00	
Male Genital System, Prostate Gland, Lymph Vessels, (Delawarr)	71.75	57.75	Freq: 20.0, 2000.0
Liver, Right Lobe, (Delawarr)	72.00	29.00	
Large Intestine, Rectum, Sacral Flexure, (Delawarr)	72.00	67.00	Freq. 10.0, 880.0
Reticulo-Endothelial System, Cells Of Bone Marrow, Reticulo-Endothelium, Plasma-, (Delawarr)	72.00	68.25	blasts
Male Genital System, Testes, Nerves, From Renal Plexus, (Delawarr)	72.25	30.50	Freq. 1.0, 880.0
Eyes, Eyeball, Fibers Of Lens, (Delawarr)	72.50	57.25	Freq: 20.0
Small Intestine, Mucous Membrane, Villi, Muscle Cells, (Delawarr)	72.50	66.00	Freq. 10.0, 880.0, 20.0, 240.0
Lymphatic System, Lymph Glands, Mesenteric, Draining Intestines, (Delawarr)	72.75	46.50	Freq. 10.0, 880.0

Patterns of Information	Rate Left	Rate Right	Energetic Processes
External & Physical Agent Disorders, Burns, Thermal, Shock, (Delawarr)	73.00	41.50	
Male Genital System, Testes, Nerves, From 10th Thoracic Segment, (Delawarr)	73.00	46.50	Freq. 1.0, 880.0
Muscular Structure, Fasciae, Synovial Bursae, (Delawarr)	73.00	87.25	
Central Nervous System, Spinal Cord, Grey Matter, Nerve Cells, (Delawarr)	73.20	46.20	
Circulatory System, Arterioles, (Delawarr)	73.25	40.75	Parts of the Arteries
Respiratory System, Trachea, Cartilages, (Delawarr)	73.25	69.00	
Small Intestine, Plexus of Minute Branches, (Delawarr)	73.50	72.00	Freq. 10.0, 880.0
Muscular Structure, Fasciae, (Delawarr)	73.50	81.50	Fascia Tissue Surrounds Organs & Muscles & Helps Keep Their Shape; Emotions Settle in these Tissues Which Tends To Lock Up the Fascia
Male Genital System, Prostate Gland, Glandular Substance, (Delawarr)	73.75	63.75	A Secretion Which is thin, opalescent, slightly alkaline fluid which forms part of the semen; Freq: Freq: 20.0, 2000.0
Ductless Glands, Pituitary Body, Anterior Lobe, (Delawarr)	74.00	46.75	With Depression Check Need To Increase Noradrenaline
Small Intestine, Arteries, Ileal, (Delawarr)	74.00	63.00	Freq. 10.0, 880.0
Ductless Glands, Pituitary Body, Anterior Lobe, (Delawarr)	74.00	65.00	With Depression Check Need To Increase Noradrenaline
Respiratory System, Lungs, Serous Coat, (Delawarr)	74.74	28.50	Freq. 9.0
Mouth, Nose & Throat, Pharynx, Nerves, Glossopharyngeal Branch, (Delawarr)	75.00	16.00	
Internal Ear, Membraneous Labyrinth, Ductus Utriculosaccularis, (Delawarr)	75.00	54.50	Freq: 20.0, 880.0
Eyes, Eyeball, Iris, Layer Of Endothelial Cells, (Delawarr)	75.00	62.50	Freq: 20.0
Male Genital System, Seminal Vesicles, (Delawarr)	75.50	63.50	
Cell Structure, Protoplasm, Cytoplasm, Golgi's Network, (Delawarr)	75.50	75.50	
Large Intestine, Lymph Vessels, (Delawarr)	75.50	80.25	Freq. 10.0, 880.0
Male Genital System, Penis, Nerves, From 3rd Sacral, (Delawarr)	75.50	86.50	
Male Genital System, Penis, Body, Glans Penis, (Delawarr)	75.50	89.00	
Muscular Structure, Tendons, Nerve Supply, (Delawarr)	75.60	69.50	
Lymphatic System, Lymph Glands, Lingual, Draining Deeper Tissues Of The Neck, (Delawarr)	76.00	35.40	Freq. 10.0, 880.0
Mouth, Nose & Throat, Pharynx, Adenoids, Adenoid Area, (Delawarr)	76.00	36.00	Freq: 20.0, 880.0
Ductless Glands, Pineal Body, Pineal Cells, (Delawarr)	76.00	47.75	Freq. 20.0, 5000.0
Internal Ear, Membraneous Labyrinth, Semicircular Ducts, Septum Transversum, (Delawarr)	76.25	73.25	Freq: 20.0, 880.0
Muscular Structure, Fasciae, Synovial Membrane, (Delawarr)	76.25	73.50	
Mouth, Nose & Throat, Mouth, Lymph Vessels, (Delawarr)	76.40	56.50	
Ductless Glands, Suprarenal (Adrenal) Glands, Fibro-areolar Capsule, (Delawarr)	76.50	71.50	
Internal Ear, Arteries of the Labyrinth, Stylo-Mastoid Branch Of Auricular Artery, (Delawarr)	76.50	76.75	Freq: 20.0, 880.0
Stomach, Helpful Paired with Spleen 4.00-95.00, (Delawarr)	77.00	84.00	Pair Color-Yellow; Emotion-Pensiveness; Sound-Singing; Freq: 20.0
Female Genital System, Vagina, Clitoris, (Delawarr)	77.10	86.10	
Male Genital System, Vas Deferens, Ductili Aberrantes, (Delawarr)	77.25	48.75	
Large Intestine, Veins, Nerves, Sympathetics, Hypogastric Plexus, (Delawarr)	77.25	62.75	Viscera in Mesenteric Inferior are Served by; Freq. 10.0, 880.0

Patterns of Information	Rate Left	Rate Right	Energetic Processes
Large Intestine, Rectum, Internal Sphincter, (Delawarr)	77.25	75.25	Freq. 10.0, 880.0
Male Genital System, Penis, Arteries, Helicine Arteries, (Delawarr)	77.50	64.00	
Large Intestine, Rectum, Perineal Flexure, (Delawarr)	77.50	77.00	Freq. 10.0, 880.0
Mouth, Nose & Throat, Salivary Glands, Parotid Gland, Nerves, Of Otic Ganglion, (Delawarr)	77.75	66.50	
Skin, Hair & Nails, Medulla Of Hair, (Delawarr)	78.50	66.00	
Skeletal System, Bone, Blood Vessels, Lacunae, (Delawarr)	78.50	80.25	
Eyes, Eyeball, Iris, Pupil, (Delawarr)	78.60	80.75	Freq: 20.0
Skeletal System, Joints, (Delawarr)	79.00	52.00	
Respiratory System, Pleurae, Nerves, (Delawarr)	79.00	69.25	
Central Nervous System, Fore Brain, Cerebral Cortex White Matter, (Delawarr)	79.00	81.00	With Depression Check Need To Increase Enkephalins, Norepinephrins, Serotonin Or Decrease Dopamine, Noradrenaline
Muscular Structure, Fasciae, Fatty Nodules, (Delawarr)	79.20	73.50	
Skin, Hair & Nails, Epidermis, Horny Zone, Horny Layer, (Delawarr)	79.25	56.50	
Large Intestine, Veins, Nerves, Parasymphathetics, Pelvic Splanchnic Nerves, (Delawarr)	79.50	68.75	Viscera in Mesenteric Inferior are Served by; Freq. 10.0, 880.0
Muscular Structure, Fasciae, Superficial Fascia, (Delawarr)	79.60	69.25	
Gall Bladder & Biliary Ducts, Gall Bladder, Veins, Cystic, (Delawarr)	80.00	37.00	
Skin, Hair & Nails, Hair, (Delawarr)	80.25	58.20	
Large Intestine, Veins, Nerves, Parasymphathetics, Vagus Branch of Large Intestine, (Delawarr)	80.25	62.00	Viscera in Mesenteric Superior are served by; Freq. 10.0, 880.0
Eyes, Eyeball, Iris, Arteries, (Delawarr)	80.25	69.00	Freq: 20.0
Central Nervous System, Brain Areas, Postrolandic Area, Trunk, (Delawarr)	80.50	63.75	
Central Nervous System, Hind Brain, (Delawarr)	80.75	72.00	Main Area of Brain
Reticulo-Endothelial System, Cells Of Bone Marrow, Yellow Marrow, Monocytes, (Delawarr)	81.10	60.20	
Circulatory System, Arteries, (Delawarr)	81.25	78.00	
Small Intestine, Mucous Membrane, Villi, Artery Leaving Villus, (Delawarr)	81.50	17.60	Freq. 10.0, 880.0
Male Genital System, Seminal Vesicles, and Ejaculatory Ducts, (Delawarr)	81.50	48.60	
Female Genital System, Uterus, (Delawarr)	82.00	22.00	
Ductless Glands, Thyroid Gland, (Delawarr)	82.00	51.00	Freq. 20.0, 160.0
Female Genital System, Ovaries, (Delawarr)	82.00	65.50	Freq. 1.0, 880.0
Male Genital System, Penis, Veins, (Delawarr)	82.00	69.50	
Male Genital System, Penis, Body, Corpora Cavernosum Penis, (Delawarr)	82.00	71.50	
Central Nervous System, Brain Areas, Sensory Area, Smell, (Delawarr)	82.20	60.50	Check Hypocampus, Bi-Polar
Female Genital System, Uterus, Cervix Of Uterus, Mucous Glands, (Delawarr)	82.50	70.00	
Skeletal System, Bone, Nerves, (Delawarr)	83.00	68.00	
Male Genital System, Prostate Gland, Interstitial Tissue, (Delawarr)	83.25	61.50	Freq: 20.0, 2000.0
Reticulo-Endothelial System, Cells Of Bone Marrow, Reticulo-Endothelium, Megakarycocytes, (Delawarr)	83.25	65.25	blasts
Internal Ear, Membraneous Labyrinth, Saccule, (Delawarr)	83.40	78.75	Freq: 20.0, 880.0
Reticulo-Endothelial System, Cells Of Bone Marrow, Reticulo-Endothelium, Thrombocytes, Platelets,, (Delawarr)	83.50	95.50	As Fragments of megakarycoytes with life-span of 5-9 days, at death give off serotonin, With Depression Check Need To Increase Norepinephrins, Or Decrease Noradrenaline
Large Intestine, Veins, Hemorrhoidal Venous Plexus, (Delawarr)	83.75	56.00	Freq. 10.0, 880.0

Patterns of Information	Rate Left	Rate Right	Energetic Processes
Male Genital System, Prostate Gland, Lateral Lobes, (Delawarr)	84.25	61.00	Freq: 20.0, 2000.0
Large Intestine, Rectum, (Delawarr)	85.00	01.00	Freq. 10.0, 880.0
Mouth, Nose & Throat, Mouth, Tongue, (Delawarr)	85.00	43.00	
Male Genital System, Prostate Gland, Glandular Tissue, (Delawarr)	85.00	56.00	Freq: 20.0, 2000.0
Large Intestine, Structure Of, Arteries, Mesenteric Inferior, Branching To Descending, (Delawarr)	85.00	66.00	Colon, Transverse Colon (Left Hypochondriac Region), Pelvic Colon, Rectum; Freq. 10.0, 880.0
Pancreas, (Delawarr)	85.50	22.50	Chromium Needs to Be In Balance for Proper Function; Freq. 15.0
Eyes, Eyeball, Fibrous Coat, Cornea, Posterior Elastic Lamina, (Delawarr)	85.50	54.40	Freq: 20.0
Respiratory System, Lungs, Intrapulmonary Bronchi, (Delawarr)	85.75	63.50	Freq. 9.0
Ductless Glands, Parathyroid Gland, (Delawarr)	86.00	04.00	If High Tendency to Clot, Balance Thyroid with Parathyroid, Located Near Or Alongside the Thyroid gland, Secretes Hormone Important In Controlling Calcium-Phosphorus Balance
Female Genital System, Uterus, Lymph Vessels, (Delawarr)	86.25	10.50	
Muscular Structure, Muscular Tissue, Cardiac, (Delawarr)	86.50	48.40	
Central Nervous System, Spinal Cord, Meninges, Pia Mater, (Delawarr)	87.00	34.00	
Circulatory System, Veins, Nerves, (Delawarr)	87.25	43.00	
Female Genital System, Uterus, Ligaments, Round Ligaments, (Delawarr)	87.25	64.75	
Respiratory System, Lungs, (Delawarr)	88.25	08.60	Afected by Lyme Disease; Freq. 9.0
Spleen & Thymus, Spleen, Left Side Under Arm, (Delawarr)	88.50	09.50	Freq: 20.0
Male Genital System, Penis, Body, Neck Of Penis, (Delawarr)	88.75	46.50	
Large Intestine, Cecum, (Delawarr)	90.00	66.00	Freq. 10.0, 880.0
Muscular Structure, Muscular Tissue, Nerve Supply, Sensory Nerves, (Delawarr)	90.25	81.00	
Large Intestine, Rectum, External Sphincter, (Delawarr)	90.25	87.00	Freq. 10.0, 880.0
Large Intestine, Rectum, Rectal Ampulla, (Delawarr)	90.50	87.50	Freq. 10.0, 880.0
Skeletal System, Bone, Blood Vessels, Canaliculi, (Delawarr)	90.75	82.40	
Skin, Hair & Nails, Epidermis, (Delawarr)	91.00	31.00	
Spleen & Thymus, Spleen, Peritoneum, (Delawarr)	91.00	31.00	Freq: 20.0
Muscular Structure, Aponeuroses, (Delawarr)	91.50	68.00	
Female Genital System, Vagina, (Delawarr)	92.00	31.00	
Respiratory System, Bronchi, (Delawarr)	92.00	82.00	
Respiratory System, Lungs, Helpful to Pair with Large Intestine 8.00-72.00, (Delawarr)	92.00	83.00	Affected by Lyme Disease, Support for Neutralize Tobacco Toxins, Pair Color-White; Emotion-Sadness; Sound-Weeping; Freq. 9.0
Gastrointestinal Disorders, Stomach, Ulcers & Sores, Pain Increased by Food, (Delawarr)	92.00	85.00	& at Night, Use 5.00-91.00 Prostate
Female Genital System, Vagina, (Delawarr)	92.00	96.00	
Reticulo-Endothelial System, Cells Of Bone Marrow, Reticulo-Endothelium, Megakary-, (Delawarr)	93.60	85.75	ocytes, Found in Red Bone marrow then Evolves into thrombocytes or blood platelets at their death give off the chemical serotonin
Respiratory System, Pleurae, Lymph Vessels, (Delawarr)	95.25	37.60	